VOLUME SEVENTY

Advances in
ECOLOGICAL RESEARCH
Stream Research in Glacier bay, Alaska From 1977–2024: Part 1

ADVANCES IN ECOLOGICAL RESEARCH

Series Editors

DAVID A. BOHAN
*Agroécologie, AgroSup Dijon, INRAE,
Université de Bourgogne Franche-Comté,
Dijon, France*

ALEX J. DUMBRELL
*School of Life Sciences,
University of Essex,
Wivenhoe Park, Colchester,
Essex, United Kingdom*

VOLUME SEVENTY

Advances in
ECOLOGICAL RESEARCH

Stream Research in Glacier bay, Alaska From 1977–2024: Part 1

Edited by

ALEX J. DUMBRELL
School of Life Sciences, University of Essex,
Wivenhoe Park, Colchester, Essex, United Kingdom

ALEXANDER M. MILNER
School of Geography, Earth and Environmental Sciences,
University of Birmingham, Edgbaston, Birmingham,
United Kingdom; Institute of Arctic Biology,
University of Alaska Fairbanks, Fairbanks, AK, United States

Academic Press is an imprint of Elsevier
125 London Wall, London, EC2Y 5AS, United Kingdom
50 Hampshire Street, 5th Floor, Cambridge, MA 02139, United States
525 B Street, Suite 1650, San Diego, CA 92101, United States

First edition 2024

Copyright © 2024 Elsevier Ltd. All rights are reserved, including those for text and data mining, AI training, and similar technologies.

Publisher's note: Elsevier takes a neutral position with respect to territorial disputes or jurisdictional claims in its published content, including in maps and institutional affiliations.

No part of this publication may be reproduced or transmitted in any form or by any means, electronic or mechanical, including photocopying, recording, or any information storage and retrieval system, without permission in writing from the publisher. Details on how to seek permission, further information about the Publisher's permissions policies and our arrangements with organizations such as the Copyright Clearance Center and the Copyright Licensing Agency, can be found at our website: www.elsevier.com/permissions.

This book and the individual contributions contained in it are protected under copyright by the Publisher (other than as may be noted herein).

Notices
Knowledge and best practice in this field are constantly changing. As new research and experience broaden our understanding, changes in research methods, professional practices, or medical treatment may become necessary.

Practitioners and researchers must always rely on their own experience and knowledge in evaluating and using any information, methods, compounds, or experiments described herein. In using such information or methods they should be mindful of their own safety and the safety of others, including parties for whom they have a professional responsibility.

To the fullest extent of the law, neither the Publisher nor the authors, contributors, or editors, assume any liability for any injury and/or damage to persons or property as a matter of products liability, negligence or otherwise, or from any use or operation of any methods, products, instructions, or ideas contained in the material herein.

ISBN: 978-0-443-29732-8
ISSN: 0065-2504

For information on all Academic Press publications
visit our website at https://www.elsevier.com/books-and-journals

Publisher: Zoe Kruze
Acquisitions Editor: Jason Mitchell
Editorial Project Manager: Dewwart Chauhan
Production Project Manager: Abdulla Sait
Cover Designer: Arumugam Kothandan

Typeset by MPS Limited, India

Contents

Contributors ix

1. Glacier Bay National Park, Alaska: A natural laboratory to study stream development and succession following deglaciation   1
Alexander M. Milner

1. Glacial history	2
2. Human settlements	3
3. Explorers and visitors	3
4. Succession	4
5. Primary successional stream studies	6
6. National Park Service	8
7. Previous river research	9
8. Outline of papers in the special issues vol 70 and 71	13
Acknowledgments	17
Funding	17
References	17

2. The spatiotemporal dynamics of the physical habitat template at a watershed scale chronosequence in Glacier Bay, southeast Alaska   23
Svein Harald Sønderland and Alexander M. Milner

1. Introduction	24
2. Methods	30
3. Results	35
4. Discussion	42
5. Summary	56
Acknowledgements	57
Appendix 1. Watershed variables correlation table	58
References	60

3. Role of riparian vegetation in colonization and succession of stream macroinvertebrates in Glacier Bay, Alaska — 67
Elizabeth Flory, Ian Gloyne-Phillips, Amanda J. Veal, and Alexander M. Milner

1. Introduction	68
2. Methods	70
3. Results	75
4. Discussion	88
Appendix	94
References	96

4. Chironomidae (Diptera) community succession in streams across a 200 year gradient in Glacier Bay National Park, Alaska, USA — 99
Alexander M. Milner, Katrina Magnusson, Amanda J. Veal, and Lee E. Brown

1. Introduction	100
2. Methods	101
3. Data analysis	105
4. Results	108
5. Discussion	113
6. Summary	115
Acknowledgments	116
Appendix	116
References	118

5. Salmon lice (*Lepeophtheirus salmonis*) as a food source for juvenile salmonids in Glacier Bay, southeast Alaska — 121
Svein Harald Sønderland and Alexander M. Milner

1. Introduction	122
2. Methods	124
3. Sample collection	125
4. Temperature and salinity measurements and pink salmon enumeration	125
5. Results	127

6. Discussion	127
7. Conclusion	133
Acknowledgments	133
References	133

6. Convergence of beta diversity in river macroinvertebrates following repeated summer floods   **137**

Lawrence J.B. Eagle, Alexander M. Milner, Megan J. Klaar, Jonathan L. Carrivick, and Lee E. Brown

1. Introduction	138
2. Study area	141
3. Datasets and methods	144
4. Results	148
5. Within site beta-diversity	151
6. Between river beta-diversity	153
7. Discussion	154
8. Conclusion	159
Acknowledgements	160
Appendix 1	160
References	162

Contributors

Lee E. Brown
School of Geography, University of Leeds, Woodhouse Lane, Leeds, United Kingdom

Jonathan L. Carrivick
School of Geography, University of Leeds, Woodhouse Lane, Leeds, United Kingdom

Lawrence J.B. Eagle
School of Geography, University of Leeds, Woodhouse Lane, Leeds, United Kingdom

Elizabeth Flory
Aquatic Science Inc., Juneau, AK, United States

Ian Gloyne-Phillips
Marine Group, NIRAS UK, Liverpool, United Kingdom

Megan J. Klaar
School of Geography, University of Leeds, Woodhouse Lane, Leeds, United Kingdom

Katrina Magnusson
School of Geography, Earth and Environmental Sciences, University of Birmingham, Edgbaston, Birmingham, United Kingdom

Alexander M. Milner
School of Geography, Earth and Environmental Sciences, University of Birmingham, Edgbaston, Birmingham, United Kingdom; Institute of Arctic Biology, University of Alaska Fairbanks, Fairbanks, AK, United States

Svein Harald Sønderland
School of Geography, Earth and Environmental Sciences, University of Birmingham, Edgbaston, Birmingham, United Kingdom; Institute of Arctic Biology, University of Alaska Fairbanks, Fairbanks, AK, United States

Amanda J. Veal
School of Geography, Earth and Environmental Sciences, University of Birmingham, Edgbaston, Birmingham, United Kingdom

CHAPTER ONE

Glacier Bay National Park, Alaska: A natural laboratory to study stream development and succession following deglaciation

Alexander M. Milner[a,b,*]

[a]School of Geography, Earth and Environmental Sciences, University of Birmingham, Edgbaston, Birmingham, United Kingdom
[b]Institute of Arctic Biology, University of Alaska Fairbanks, Fairbanks, AK, United States
*Corresponding author. e-mail address: a.m.milner@bham.ac.uk

Contents

1. Glacial history	2
2. Human settlements	3
3. Explorers and visitors	3
4. Succession	4
5. Primary successional stream studies	6
6. National Park Service	8
7. Previous river research	9
8. Outline of papers in the special issues vol 70 and 71	13
Acknowledgments	17
Funding	17
References	17

Abstract

Ecosystems at the glacier margins can serve as natural laboratories to research climate change impacts. Although recession following the recent ice advance was not due to climate change, the same premise still applies. Here we look at the glacial history of Glacier Bay dating back to 18,000 years Before Present, including the last glacial advance that filled almost the entire bay reaching its maximum in 1780. We look at the human settlement in Glacier Bay, including potential settlements when the ice retreated. Visitors and explorers are looked at since the ice retreated and an overview of the terrestrial plant succession is subsequently provided. Glacier Bay is touted as classic facilitation in many Ecology textbooks, but this facilitation often is not the case as laid out. The next section in this chapter is an overview of the stream successional work that has been carried out over the last 40+ years and outlines the main findings. Then there is an overview of each paper in the special issue(s), with the title, authors, and the content.

1. Glacial history

The Northeast Gulf of Alaska coast, including the Glacier Bay area, is one of the most intensely glaciated portions of the northern hemisphere, producing the extreme, insular present landscape. There is evidence for ice at sea level by 10 million years Before Present (BP) and as recently as 18,000 BP, virtually the entire region was icebound (Mann & Gaglioti, 2024). Ice retreated from that maximum to the inner fjords of the Coast Mountains by 14,000 BP (Mann & Hamilton, 1995). Scattered evidence suggests that ice was not extensive in Glacier Bay proper from the Younger Dryas cold period, c. 12–11,000 BP, through the mid Holocene, probably reaching low elevations only in the upper west arm of the Bay (Mann & Gaglioti, 2024). By 5500BP West Arm ice in Glacier Bay extended south to impound a lake in the largely ice free Muir Inlet lowlands (Goodwin, 1988). After 2520 years BP the West Arm retreated, draining "Muir Lake" (Connor et al., 2009). There was a readvance about 1600–1000BP when West Arm ice and outwash again blocked lower Muir Inlet sufficiently to create a "Glacial Lake Adams", but after 1200BP that lake was extinguished by a combination of glacial and sedimentary activity in the Muir inlet basin. (Connor et al., 2009; Wiles et al., 2011). This period of episodic advance probably was intensified around 1000BP during a global cold period (Gaglioti et al., 2019). Ice advancing out of the West Arm and Muir Inlet coalesced and by 500BP their joint terminus lay in the middle reaches of Glacier Bay, delimited by a large terrestrial floodplain (Mann & Streveler, 2008; Connor et al., 2009). Ice rapidly advanced thereafter, its maximum modern ice extent occurring around 1780; (Connor et al., 2009; Wiles et al., 2011). At its maximum, the terminus lay in Icy Strait and its icefield covered >6000 km^2 and thickness was up to 1.5 km (Larsen et al., 2005).

After the late Neoglacial maximum the terminus retreated rapidly, reaching the heads of Tarr and Johns Hopkins Inlets by 1916 AD; (Mann & Ugolini, 1985). Ice front retreat in the East Arm began later, about 1900 AD and continued to the head of Muir Inlet until 1992 when the Muir Glacier terminus retreated onto land (Hunter et al., 1996; Molnia, 2007). A marine terminus persists in McBride Inlet to this day. Elsewhere in the Bay, tidewater termini persist only in upper Tarr and Johns Hopkins inlets, while a third, Lamplugh Glacier hangs on tenuously. Abrupt ice removal in Glacier Bay has resulted in very large isostatic rebound rates. Most sectors are uplifting at 10 mm yr^{-1} but several sites are uplifting more rapidly at 25 mm yr^{-1} (Larsen et al., 2004). The postglacial Glacier Bay landscape of today is dominated by scoured bedrock features, though remnant outwash surfaces are important locally, especially in the Muir Inlet and lower bay areas.

There were areas of Glacier Bay National Park's outer coast and associated continental shelf that may have been used as ice-free refugia during the Late Glacial Maximum. Though Arctic in character and thus of limited use as temperate refugia (Mann, 1983), they may have served as stepping stones for early maritime humans entering the western hemisphere from Asia (Carrara et al., 2007). Present day populations of chum salmon are likely derived from late Pleistocene populations that existed in refugia stream channels on the outer coast which are now underwater (Kondzela et al., 1994).

The climate of Glacier Bay is maritime, with mean annual air temperature of 5°C (mean monthly range of −3°C to 13°C) and average annual precipitation of 1400 mm (Milner & Robertson, 2010).

2. Human settlements

The indigenous people of northern SE Alaska, the Huna Tlingit's oral history indicates that they used to populate southern parts of Glacier Bay before the LIA reached it maximum (Connor et al., 2009) and they had to flee to various points along Icy Strait. Oral history describes a pre-advance broad valley and village with a meadow-lined river flowing along its western margin called the Grassy River (Chookanheeni) from which the Chookanheeni Clan takes its name. When the LIA ice retreated, the Tlingit consider a stream in the northwest corner of Berg Bay to be the modern manifestation of the Chookanheeni River (Monteith, et al. 2007). There were two other rivers on the east side of the Bay, the Ghatheeni (Sockeye River) and the Ghatheeni Tlein (Big Sockeye River)—the analogous present day rivers are the Bartlett and Beartrack Rivers (Monteith, 2017). A second, larger Sand Mountain Village was located in the south-central part of the plain in association with one of these rivers (Connor et al., 2009).

3. Explorers and visitors

When George Vancouver sailed in Icy Strait in 1794 the recession of the Glacier Bay icesheet had started as the ice was thought to be at the

Beardslee Islands and was discharging icebergs (Cooper, 1923). When John Muir visited for the first time in 1879 (the dominant east arm glacier was named after him), Muir Glacier was separate from the glacier in the northwest arm. He returned in 1880 and 1890 establishing his main base of operations close to Muir Glacier then at the base of Mt. Wright (Cooper, 1923). Since his first appearance 1879 there has been unbroken history of scientific study with the glaciers being studied and then from 1916, when WS Cooper first visited, terrestrial ecological patterns have been consistently studied (Rumore, 2009).

Tourists started visiting Glacier Bay in 1890s from writings of John Muir when the steamer "Queen", "Idaho" and others began a long series of tourist excursions to Muir Glacier—but excursions came to an end in 1899 when rapid glacial recession created enormous collections of floating ice dangerous for ships (Cooper, 1923). The Ecological Society of America, at the behest of William Cooper, petitioned the federal government to make Glacier Bay a National Monument which President Coolidge did so in 1925. In 1980, President Jimmy Carter signed the Alaska National Interest Lands Conservation Act creating Glacier Bay National Park and Preserve (GBNP). The area fosters unique opportunities for scientific studies of tidewater glacial landscapes and associated natural successional processes and preserves one of the largest units of the national wilderness preservation system, encompassing more than 1.1 million ha of glacially influenced marine, terrestrial, and freshwater ecosystems (Capps, 2017).

Importantly from an ecological perspective, most tourists visit on cruise ships or the day boat so not interfering with successional processes, particularly terrestrial processes. Over 50% of visitors said the most important experience was seeing wildlife, whereas only 31% was seeing the glacier (Furr, 2019).

4. Succession

Succession is one of the oldest concepts in ecology (Pickett, 1982); its study has generated mechanisms by which communities are assembled including the relative role of deterministic and stochastic processes (Milner & Robertson, 2010). Deterministic communities follow predictable trajectories whereas stochastic communities are formed by chance colonization and extinction and is non-directional (Lepori & Malmqvist, 2009). The succession in the main part of Glacier Bay is primary succession as no remnants of the previous community remains when the ice retreats. Connell and Slatyer (1977) proposed three models of successional processes; (1) facilitation where early species modify the environment to make it more suitable for later species, (2) inhibition where early colonizers restrict the invasion of later colonizers, and (3) tolerance where a progressive tolerance of invading species occurs with the changing environmental conditions and few species are lost. In all the general Ecology textbooks after deglaciation, terrestrial succession is thought to occur by facilitation of later colonizers by earlier plant species (Cooper, 1923; Lawrence, 1958). The classic terrestrial succession in Glacier Bay is N fixing mountain avens (*Dryas drummondii*) and scattered willows (*Salix* spp.) among lichens, liverworts, horsetail after 25 years which gives way to N fixing alder (*Alnus* spp.) which dominates the landscape after 50 years. The nutrients in the soil facilitate a later colonizer Sitka spruce (*Picea sitchensis*) which dominates after 100 years and then increasing amounts of Western hemlock (*Tsuga heterophylla*).

However Chapin et al. (1994) suggest that inhibition may also occur in plant primary successional processes and spruce can establish without alder, particularly close to a seed source. Three of the older sites Fastie (1995) studied were distinguished by the rarity of spruce trees with a distinctive of suppression which was present >75% of spruce at younger sites. Thus, the plant communities at sites of different ages do not constitute a single chronosequence, and should not be used to infer long term successional trends as there may be multiple pathways at play (Fastie, 1995).

Buma et al. (2019) occupied Cooper's original plots which were mostly in the west arm of Glacier Bay and found that they do not support the classic facilitation model of succession and stochastic early community and subsequent inhibition have dominated just as Fastie (1995) found in lower Glacier Bay. Age of the site provided less predictive of community composition than spatial location (Buma et al., 2017). Thus the potential limitations raised by Johnson and Miyanishi (2008) of using a chronososeequence approach to inferring community change and development are valid. The chronosequence approach is appropriate for communities that change in a linear fashion at a landscape level or inferring soil organic matter or soil nutrients (Walker et al., 2010).

Wetlands and peatlands are rare on Neoglacial landscape (Boggs et al., 2010) due to the generally late-successional nature of bogs, due to their relationship to paludification and isolation by peat accumulation from bedrock.

5. Primary successional stream studies

The stream studies started in the summer of 1977 when a Chelsea College expedition of eight (5 freshwater scientists, two cameramen and a cook!) visited the Park (Fig. 1) and sampled several sites in nine streams—some streams were flowing directly from remnant ice. We also examined about 6 lakes and kettle ponds. This was the start of a long history of research into lotic successional processes that continues to present day (see below). Initially the research focused on benthic invertebrates in the streams and their driving variables and relationship to watershed age. Then the research focused on different aspects of the lotic system; role of riparian vegetation, instream competition, effects of spawning anadromous salmon and diet of juvenile salmonids rearing in the stream, role of coarse woody debris in providing habitat, role of disturbance for example floods resetting successional processes and eliminating later stream colonizers, and studies of the interactions between the terrestrial ecosystem and the stream ecosystem.

However, it is difficult to conduct stream research in remote places, especially when you cannot drive to the stream. Thus, it needs boat transportation to get to the streams, as acknowledged below. Initially the work by the PhD students involved a base camp, with a wall tent over a wood floor and frame and included a wood stove (Fig. 2). The locations were Nunatak Cove (1978–80) and Goose Cove (1990–98) both camps for Wolf Point Creek and then Stonefly Creek (1999–2005) as the main research focus were on those creeks. Subsequently, there were no base camps and PhD student research involved a number of stream systems with different watershed ages and research students camped near the creek they were sampling for short periods. Nevertheless, the longest-term record was Wolf Point Creek, where macroinvertebrate samples were collected almost every year (Fig. 3).

Of Connell and Slatyer (1977)'s three mechanisms of succession, the one mechanism that applies most to stream succession in Glacier Bay is tolerance as the majority of macroinvertebrates remain in the system after colonization by other taxa unless they are poor competitors (e.g. the chironomid *Diamesa*) (Milner et al., 2011; Milner et al., 2008). Only when major disturbances in stream systems occur can these poor competitors recolonize as their main invertebrate competitor is reduced to low numbers. Tolerance is rarely shown in other ecosystems, either being inhibition or facilitation.

One aspect of the stream ecosystem is that successional processes do not occur in isolation of other ecosystems—terrestrial, lake or the marine

Fig. 1 Chelsea College expedition to Glacier Bay near Wolf Point Creek - the cook is first on the left.

Fig. 2 Base camp NR Stonefly Creek equipped with a stove, propane burner and light.

intertidal. We tried to encapsule these processes in a BioScience paper (Milner et al., 2007) where the strengths of links between the different ecosystems changes during different time periods since deglaciation—we used 0–5 years, 5–50 years, 50–150 years and 150+ years (Fig. 1 from the publication and see Klaar et al., this issue). Initially in the first 2 time periods abiotic processes dominate but then biotic processes have more of a

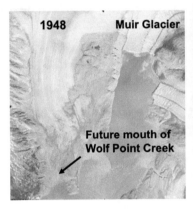 

Fig. 3 Wolf Point Creek in 1948 and today.

role in linkages >50 years, but the extent of the biotic control varies among ecosystems. Change in one ecosystem has a major influence on the nature and direction of change in the other ecosystems.

6. National Park Service

National Park Service (NPS) facilitated the research over the years, mainly with logistical and manpower assistance. We especially acknowledge the assistance of Jim Luthy of the *mv Nunatak* and Justin Smith and Todd Bruno of the *mv Capelin* who transported material for the base camps. Over the many years there were 7 Superintendents of GBNP governing the National Park and 8 Resource Managers to liaise with, particularly the latter. One aspect you realize over the years is that the Superintendent governs the Park, and every Superintendent acts differently. There is no governance at the regional level—it is all at the Park level. Only two of these Superintendents came in the field with us. Initially in the late 1970s two people were in Resource Management which grew to >10 in the 2010s and consequently regulations increased. One particular concern was the motorless Wachusett Inlet between 15 July and end of August (there was no scientific input into this discussion)—we always tried to get an exception as we needed to count adult fish spawners and sample invertebrates at this time of the year in Stonefly Creek at the end of Wachusett Inlet. Usually, we were successful, even though one or two disputes occurred.

7. Previous river research

As the Neoglacial ice was ponded in Muir Inlet by the West Arm ice, the vast glaciofluvial/lacustrine deposits accumulated in the Muir Basin during the Holocene was less eroded (Greg Streveler pers. comm.) These remaining deposits form the basis of the watersheds that we have studied. Another aspect of these deposits in the watersheds was the Beardslee Formation where fluvial deposits partially escaped ice erosion due to the rapidity of the advance/retreat of the final ice surge in the lower bay (Connor et al., 2009).

This is a brief summary of the findings of the previous river research that has been accomplished over the past 40+ years. Initially the findings were a comparison of macroinvertebrates in nine streams, in light of the rapid recession of a Neoglacial ice sheet within the last 250 years. The Chironomidae (Diptera) were the pioneer invertebrate colonizers of newly emergent streams arising as meltwater from receding ice sheets and displayed a distinct pattern of succession with stream maturity. Environmental variables shown to be most significant in stream development were water temperature and channel stability (Milner, 1987). The primary factors governing the salmonid establishment, species diversity composition and abundance of salmonids in Glacier Bay streams were water temperature, sediment loading and stream discharge. No salmonids were found in the turbid meltwater streams emerging from retreating ice up to 1988. Dolly Varden (*Salvelinus malma* Walbaum) charr were the first salmonids to colonize the youngest clearwater stream (Milner & Bailey, 1989). Milner (1994) studied macroinvertebrate colonization of a new stream Wolf Point Creek for a 12 year period from 1978 to 1990. Invertebrates, particularly Chironomidae, displayed site-specific temporal succession over the study period, at the end of which a number of the pioneer colonizers were no longer collected. Maximum species richness was found in 1988, whereas total invertebrate abundance was greatest in 1978. Deterministic trends were apparent in patterns of invertebrate colonization and succession, especially the variable water temperature (Milner, 1994). These data were used for a model of Glacial Rivers (Milner & Petts, 1994; Milner, 2016). Species richness in August increased to 16 taxa in 1994 and taxa, never previously collected, colonized the stream between 1991 and 1994 including representatives of Coleoptera, Muscidae, Trichoptera, and the first non-insect taxon, Oligochaeta, in 1994 (Flory & Milner, 2000).

Flory and Milner (1995) showed that pioneer species (e.g. *Diamesa*) in rivers were eliminated by competition by later colonizers and not water temperature in experimental setups (Fig. 4). These authors also showed that the development of riparian vegetation on benthic macroinvertebrate assemblages, including their use of willow catkins entering streams in the summer and submerged alder roots as a substrate for attachment and as a source of building material for caddisfly cases. Thus riparian vegetation therefore plays an important role in the successional sequence of macroinvertebrates and overall assemblage development in Wolf Point Creek (Flory & Milner, 1999). Milner and Gloyne-Phillips (2005) in a study of 16 streams showed the importance of trailing riparian habitat for macroinvertebrates and obligate invertebrates were found on large woody debris that had fallen into older systems.

Meiofauna arthropod community structure research was started in 1998 when five streams were examined and meiofauna comprised species of Cladocera, Copepoda, Ostracoda and Hydrachnid (Robertson & Milner, 1999; Robertson & Milner, 2006). Initial colonization was rapid for meiofauna and clear evidence of successional patterns occurred, at both the species and the community level. Meiobenthic arthropod species richness and density increased along the chronosequence and was correlated with increasing stream age, coarse benthic organic matter (CBOM) and channel stability. Meiofauna colonization may have been facilitated by the large

Fig. 4 Experimental set ups and getting them there in Wolf Point Creek to test whether competition was a major influence.

breeding colonies of Canada geese that occur on the wetlands associated with newly deglaciated areas of Glacier Bay, as with the many macroinvertebrate colonizers of new streams (Robertson & Milner, 1999; Robertson & Milner, 2001) showed in Wolf Point Creek that CBOM increased the abundance and diversity of meiofauna and as a substrate to live on.

In May 1997, physical and biological variables were studied in 16 streams of different ages and contrasting stages of development following glacial recession increasing the number of streams, many for the first time (Milner et al., 2000). This sampling of these additional streams has not been repeated. The number of microcrustacean and macroinvertebrate taxa and juvenile fish abundance and diversity were significantly greater in older streams. Microcrustacean diversity was related to the amount of instream wood and percent pool habitat, while the number of macroinvertebrate taxa was related to bed stability, amount of instream wood, and percent pool habitat. Juvenile Dolly Varden (*S. malma*) were dominant in the younger streams, but juvenile coho salmon (*Oncorhynchus kisutch*) abundance was greater in older streams associated with increased pool habitat. Upstream lakes significantly influenced channel stability, percent Chironomidae, total macroinvertebrate and meiofaunal abundance, and percent fish cover. Stable isotope analyses indicated nitrogen enrichment from marine sources in macroinvertebrates and juvenile fish in older streams with established salmon runs (Milner et al., 2000).

While continuing the long time record at Wolf Point Cree; Monaghan and Milner (2009) showed that spawning pink salmon in odd years (1997) created a major disturbance by redd digging that reduced the diversity of macroinvertebrate taxa and certain fugitive taxa (e.g. Simuliidae) increased in proportional abundance (Milner et al., 2008). Colonization by large numbers of pink salmon bring marine derived nitrogen into the system and exposed decaying flesh provides a food source for macroinvertebrates (Monaghan & Milner, 2008b; Monaghan & Milner, 2008a).

Klaar et al. (2009) showed geomorphic and hydraulic complexity increased as stream age increased. Younger streams were dominated by fast flowing geomorphic units, such as rapids and riffles, with little hydraulic or landscape diversity. As streams age, slower flowing habitat units such as glides and pools became more dominant, resulting in increased geomorphic, hydraulic and landscape diversity favored by instream biota, thereby enhancing biodiversity and abundance. Also Klaar et al. (2011) showed wood characteristics altered with watershed age as terrestrial succession progressed and wood was recruited into the riverine environment. Findings showed that the size, complexity, and orientation of wood accumulations are the main drivers in determining the degree of influence instream wood have on stream geomorphic and hydraulic complexity. Adjacent terrestrial vegetation must be of a sufficient stage of development (in terms of size and maturity) to elicit significant hydrogeomorphic changes to benefit aquatic biota such as fish, macroinvertebrates, and plants.

There was a massive rain on snow flood event in November 2005, which we examined the effect of intense disturbance on the macroinvertebrate community in Wolf Point Creek (Milner et al., 2013) (Fig. 5). Pink salmon were reduced to one-tenth of pre-flood spawner densities in 2007 due to the washout of eggs in November 2005 but recovered within two generations in 2011. Macroinvertebrate community structure was significantly different after the flood as some pioneer taxa, which had become locally extinct, recolonized whereas some later colonizers (e.g. Dytiscidae beetles, Planorbidae snails, Gammaridae and Corixidae water boatmen were eliminated (Milner et al., 2018). The trajectory of the macroinvertebrate succession was reset towards a community structure of 15 years earlier. Meiofaunal abundance recovered rapidly, and richness increased post-flood with some previously unrecorded taxa colonizing. This disturbance contrasted with repeated floods in the summer of 2014 (Milner et al., 2018). The macroinvertebrate community had not

Fig. 5 Photographs of Wolf Point Creek before and after the 2005 November flood—note the stream was cut to 50% of its former width.

recovered to the preflood state when recurrent summer flooding reset the ecosystem back to an even earlier macroinvertebrate community. Total macroinvertebrate density was reduced in the winter flood by an order of magnitude more than the summer flood (Eagle et al., 2021). Meiofaunal invertebrates were more resilient to the flooding than macroinvertebrates, possibly due to their smaller body size facilitating greater access to in-stream refugia. Pacific pink salmon escapement was not markedly affected by the summer flooding, which occurred before the majority of the salmon eggs had been laid in the gravels.

In Malone et al. (2018) we used a two-century, postglacial chronosequence in Glacier Bay, Alaska, to explore the influence of C and N dynamics on soil and leaf stable isotopes. Carbon dynamics were closely linked to soil hydrology, with increasing soil water retention during ecosystem development resulting in a linear decrease in foliar and soil delta C-13, independent of shifts in vegetation cover and despite constant precipitation across sites. N dynamics responded to interactions among soil development, vegetation type, microbial activity, and topography.

8. Outline of papers in the special issues vol 70 and 71

Apart from this introductory paper, the first subsequent paper is entitled "The spatiotemporal dynamics of the physical habitat template at a watershed scale chronosequence in Glacier Bay, southeast Alaska" is by Sønderland and Milner. The paper uses GIS to characterize the five long term stream systems we have looked at intensively and in that paper gives the main characteristics of these watersheds including today's age since the stream mouth was uncovered from the glacial ice. The paper provides a map of the five watersheds and five figures with the 5 different watersheds accounting for the following different variables: (1) stream order, (2) vegetation cover and height, (3) watershed slope, (4) topographic wetness index and (5) 25 m contour lines with waterbodies. Based on the characteristics found over the chronosequence, a physical habitat template on a watershed scale at a point in time is proposed, and the directional development through deterministic chaos and noise (stochastic) events.

A paper by Flory et al. entitled "Role of riparian vegetation in colonization and succession of stream macroinvertebrates in Glacier

Bay, Alaska". Willow catkins have a major role in the colonization of certain macroinvertebrate taxa, particularly caddisflies. An experiment was undertaken to whether the growth of caddisflies was enhanced when fed by willow catkins compared to caddisflies that had algae on the substrate. The latter showed a significant weight loss. Gross primary productivity was also estimated for Wolf Point Creek.

A paper by Milner et al. is entitled "Chironomidae (Diptera) community succession in streams across a 200 year gradient in Glacier Bay National Park, Alaska, USA". This paper examines 15 streams of contrasting stages of watershed development following glacial recession were investigated in 1997 to examine chironomid community successional patterns in relation to physicochemical variables. A total of 54 taxa in five sub-families were identified; Diamesinae dominated the younger streams and Orthocladiinae in older streams. Chironominae and Tanypodinae were typically collected only in older streams. Canonical Correspondence Analysis with species and habitat data identified four distinct stream groupings (streams <50 years with lake influence; 50–100 years; 100–150 years and >150 years) based upon their chironomid community composition. CBOM, stream age and presence/absence of lakes in upper reaches of streams were the most significant habitat variables influencing communities. Total abundance and taxa richness decreased with stream age, whereas dominance, diversity and taxon richness were similar across the streams. Changes in chironomid communities over time exhibited similarities to chironomid succession found spatially along river continuums.

There is a paper by Sønderland and Milner entitled "Salmon lice (*Lepeophtheirus salmonis*) as a food source for juvenile salmonids in Glacier Bay, southeast Alaska". To our knowledge, this is the first-time salmon lice have been documented in the diets of stream-dwelling juvenile salmonids, suggesting another route of marine derived nutrients into freshwater food webs. Salmon lice are an obligate marine macroparasite commonly found on Pacific salmonids. The abundance of salmon lice in the diet of juvenile coho salmon and juvenile Dolly Varden was significantly correlated with salinity in the surface waters at 1 m in GBNP. Abundance of adult pink salmon, which typically return in larger numbers than other species particularly in odd years, seem to be the most important vector for salmon lice reaching freshwater systems.

A further paper entitled "Convergence of beta diversity in river macroinvertebrates following repeated summer floods" is by Eagle et al. This paper will extend our current understanding of community response to high-frequency floods in the summer of 2014 in numerous ways and temporally extend previous analyses of high-frequency flood impacts to include community reassembly over multiple years. Further, it will compare the extent to which immediate impact and post-flood response are shared between rivers. Beyond this, by analysing between river beta diversity, the paper will progress our understanding of community scale response to floods at larger spatial scales. This new knowledge will provide the first high level insight into landscape scale patterns in response, both immediately following floods and during subsequent macroinvertebrate community reassembly.

A paper by Klaar et al. entitled "Instream wood acts as an ecosystem engineer in river ecosystem development following recent deglaciation". This paper progresses to the formation of a conceptual model of the role of instream wood in the development of biocomplexity and linkages between terrestrial and aquatic ecosystems. This conceptual model builds upon the work of Milner et al. (2007), which highlighted the major patterns of landscape change which occur over time within Glacier Bay between 4 ecosystems. The model highlights the importance of instream wood in driving and maintaining a number of biotic and abiotic linkages between terrestrial and aquatic ecosystems and supports the fluvial biogeomorphic succession model. The work here, however, provides an important timescale of the development of the vegetation- hydrogeomorphic feedbacks using the space-for-time substitution present in Glacier Bay. This work provides important insight into using instream wood in river restoration efforts for degraded river systems.

A paper by Monaghan and Milner is titled "Disturbance and successional processes from spawning salmon and floods in Wolf Point Creek, Glacier Bay, Alaska". This paper focuses solely on Wolf Point Creek and compares disturbances by redd digging by pink salmon and disturbances by floods, occurring once in November 2005 and repeated summer floods of 2014. These disturbance data portray succession as a spatio-temporal process of ecological differentiation where intrinsic and extrinsic community disturbance promotes development of an expanding spectrum of habitats within a dynamically integrated system. This understanding emphasizes the importance of heterogeneity and connectivity in underpinning ecological resilience against global change and highlights the potential catalytic role of salmon for ecosystem restoration.

A paper by *Clitherow et al. entitled "Variation in juvenile fish diet across a gradient of stream development in Glacier Bay, Alaska" looks at diet of juvenile salmonids in streams of different ages. Reciprocal subsidies between the riparian vegetation and the stream have been well studied in many systems over short timescales (e.g., annually). Understanding these linkages change over longer time scales, for example multi-decadal to centennial, is a significant gap in our knowledge base, but is critically important as it is over these timescales that landscape and successional processes operate. Gut contents and stable isotope analyses for juvenile fish were completed for six stream ecosystems across floodplains of different ages (since deglaciation), ranging from 46 to 213 years. The balance of terrestrial and aquatic dietary contributions for juvenile fish varied across these six streams with a higher proportion of terrestrial prey items consumed and dietary breadth in older streams. The patterns in dietary contributions were influenced by physical habitat complexity in mid-aged stream and the direct and indirect drivers of instream and riparian habitat structure.

A paper by Finn et al. entitled "Predator diet breadth and population stability of juvenile coho salmon along a habitat age gradient". The hypothesis is that heterogeneous habitats lead to increased intra-population diet breadth in generalist top predators, which then leads to more healthy and long-term stable populations. The most heterogeneous habitats have been shown to occur in mid-aged streams. Three questions were addressed by this study; (1) Do juvenile coho populations occupying intermediate-aged streams have the largest diet breadth?, (2) Do populations in intermediate-aged streams have the greatest genetic diversity?; and (3) How quickly is the signal of the recent founder event lost?

One of the final papers is by Windsor et al. entitled "Community trait composition over stream development in Glacier Bay, Alaska". Meiofauna, macroinvertebrate and fish abundance data have been collected across 11 streams along a gradient of glacial meltwater contribution of 0%–70%. The aim was to understand the relative influence of biological traits on community assembly and composition along a gradient of stream development. The study addressed 3 hypotheses (1) Streams early in their development are dominated by taxa with r-selected traits (small body size, high voltinism, short lives), (2) Relationships between traits and stream development are preserved across different taxonomic groups (meiofauna, macroinvertebrates and fish) and (3) Biological traits have a strong influence on the structure of macroinvertebrate meta-communities across developing streams.

The final paper is entitled "Looking forward; a synthesis of stream research undertaken in Glacier Bay" by Milner et al. which outlines the main take home messages after 40+ years of stream succession studies in Glacier Bay. Proposed models will be discussed including the main driving variables from filters at different ages of streams—50, 100, 150 and 200 years. The future stream development will be outlined including the physical habitat simulation model and role of climate change on new systems emerging. The limitations of the studies over 40 years and future research priorities will be discussed.

Acknowledgments

AM wishes to thank all the PhD students, postdocs and collaborators over the years who have contributed immensely to stream research from 1991 to the present time and contributed to this paper. I thank the NPS for facilitating the research and providing logistical assistance with getting to the streams and the various field assistants over the years that helped with the research. I also wish to thank Greg Streveler and Wayne Howell for comments on certain sections of this manuscript.

Funding

Initially after Alexander Milner's (AM) PhD, he lived in Juneau, Alaska so it was relatively simple to sample the streams in Glacier Bay even though no funds were available to continue the sampling apart from NPS boat transportation. AM recognized that the macroinvertebrate sampling needed to be continued and when he moved to the University of Alaska at Anchorage trips to southeast continued. It was when the first PhD student Liz Flory based at the University of Stirling in Scotland funded by the National Environmental Research Council (NERC) that research really started to roll. When AM went to the University of Birmingham UK that is when stream research in Glacier Bay really took off, funded mostly by NERC and various bursaries for PhD students (10 in total). So, the motto is, even if virtually no funding, keep the long-term record going as we did for 12 years. Long-term continual datasets are rare in stream ecology.

References

Boggs, K., Klein, S.C., Grunblatt, J., Boucher, T., Koltun, B., Sturdy, M., et al., 2010. Alpine and subalpine vegetation chronosequences following deglaciation in Coastal Alaska. Arct. Antarct. Alp. Res. 42, 385–395.

Buma, B., Bisbing, S., Krapek, J., Wright, G., 2017. A foundation of ecology rediscovered: 100 years of succession on the William S. Cooper plots in Glacier Bay, Alaska. Ecology 98, 1513–1523.

Buma, B., Bisbing, S.M., Wiles, G., Bidlack, A.L., 2019. 100 yr of primary succession highlights stochasticity and competition driving community establishment and stability. Ecology 100, 100.

Capps, D.M., 2017. The role of glaciers and glacier research in the development of US National Parks. Earth Sci. History 36, 337–358.

Carrara, P.E., Ager, T.A., Baichtal, J.F., 2007. Possible refugia in the Alexander Archipelago of southeastern Alaska during the late Wisconsin glaciation. Can. J. Earth Sci. 44, 229–244.
Chapin, F.S., Walker, L.R., Fastie, C.L., Sharman, L.C., 1994. Mechanisms of primary succession following deglaciation at Glacier Bay, Alaska. Ecol. Monogr. 64, 149–175.
Clitherow, L., Sonderland, S.H., Windsor, F., Milner, A.M. Variation in juvenile fish diet across a gradient of stream development in Glacier Bay, Alaska. Adv. Ecol. Res. 71, next issue.
Connell, J.H., Slatyer, R.O., 1977. Mechanisms of succession in natural communities and their role in community stability and organization. Am. Naturalist 111, 1119–1144.
Connor, C., Streveler, G., Post, A., Monteith, D., Howell, W., 2009. The Neoglacial landscape and human history of Glacier Bay, Glacier Bay National Park and Preserve, southeast Alaska, USA. Holocene 19, 381–393.
Cooper, W.S., 1923. The recent ecological history of Glacier Bay, Alaska: II. The present vegetation cycle. Ecology 4, 223–246.
Eagle, L.J.B., Milner, A.M., Klaar, M.J., Carrivick, J.L., Wilkes, M., Brown, L., 2024. Convergence of beta diversity in river macroinvertebrates following repeated summer floods. Adv. Ecol. Res. 70, 137–169. https://doi.org/10.1016/bs.aecr.2024.09.003.
Eagle, L.J.B., Milner, A.M., Klaar, M.J., Carrivick, J.L., Wilkes, M., Brown, L.E., 2021. Extreme flood disturbance effects on multiple dimensions of river invertebrate community stability. J. Anim. Ecol. 90, 2135–2146.
Fastie, C.L., 1995. Causes and ecosystem consequences of multiple pathways of primary succession at Glacier Bay, Alaska. Ecology 76, 1899–1916.
Finn, D.S., Sonderland, S.H., Milner, A.M. Diet breadth and genetic diversity in Glacier Bay salmon. Adv. Ecol. Res. 71, next issue.
Flory, E.A., Milner, A.M., 1995. The role of competition in invertebrate community development in a recently formed stream in Glacier Bay National Park. Aquat. Ecol. 33, 175–184.
Flory, E.A., Milner, A.M., 1999. Influence of riparian vegetation on invertebrate assemblages in a recently formed stream in Glacier Bay National Park, Alaska. J. North. Am. Benthol. Soc. 18, 261–273.
Flory, E.A., Milner, A.M., 2000. Macroinvertebrate community succession in Wolf Point Creek, Glacier Bay National Park, Alaska. Freshw. Biol. 44, 465–480.
Flory, E.A., Phillips, I., Veal, A.J., Milner, A.M., 2024. Role of riparian vegetation in colonization and succession of stream macroinvertebrates in Glacier Bay, Alaska. Adv. Ecol. Res. 70, 67–99. https://doi.org/10.1016/bs.aecr.2024.09.004.
Furr, G. 2019. Current and historic visitor experiences in Coastal Alaskan wilderness: visitor motivations and experience quality in Glacier Bay National Park and Preserve.
Gaglioti, B.V., Mann, D.H., Wiles, G.C., Jones, B.M., Charlton, J., Wiesenberg, N., et al., 2019. Timing and potential causes of 19th-century glacier advances in Coastal Alaska based on tree-ring dating and historical accounts. Front. Earth Sci. 7.
Goodwin, R.G., 1988. Holocene glaciolacustrine sedimentation in Muir Inlet and ice advance in Glacier Bay, Alaska, USA. Arct. Alp. Res. 20, 55–69.
Hunter, L.E., Powell, R.D., Lawson, D.E., 1996. Flux of debris transported by ice at three Alaskan tidewater glaciers. J. Glaciol. 42, 123–135.
Johnson, E.A., Miyanishi, K., 2008. Testing the assumptions of chronosequences in succession. Ecol. Lett. 11, 419–431.
Klaar, M.J., Clitherow, L., Smith, M.W., Maddock, I., Gloyne-Phillips, I., Milner, A.M., Instream wood functions as an ecosystem engineer in river ecosystem development following recent deglaciation. Adv. Ecol. Res. 71, next issue.
Klaar, M.J., Hill, D.F., Maddock, I., Milner, A.M., 2011. Interactions between instream wood and hydrogeomorphic development within recently deglaciated streams in Glacier Bay National Park, Alaska. Geomorphology 130, 208–220.

Klaar, M.J., Maddock, I., Milner, A.M., 2009. The development of hydraulic and geomorphic complexity in recently formed streams in Glacier Bay National Park, Alaska. River Res. Appl. 25, 1331–1338.

Kondzela, C.M., Guthrie, C.M., Hawkins, S.L., Russell, C.D., Helle, J.H., Gharrett, A.J., 1994. Genetic-relationships among chum salmon populations in southeast Alaska and northern British Columbia. Can. J. Fish. Aquat. Sci. 51, 50–64.

Larsen, C.F., Motyka, R.J., Freymueller, J.T., Echelmeyer, K.A., Ivins, E.R., 2004. Rapid uplift of southern Alaska caused by recent ice loss. Geophys. J. Int. 158, 1118–1133.

Larsen, C.F., Motyka, R.J., Freymueller, J.T., Echelmeyer, K.A., Ivins, E.R., 2005. Rapid viscoelastic uplift in southeast Alaska caused by post-Little Ice Age glacial retreat. Earth Planet. Sci. Lett. 237, 548–560.

Lawrence, D.B., 1958. Glaciers and vegetation in south-eastern Alaska. Am. Scientist 46, 89–122.

Lepori, F., Malmqvist, B., 2009. Deterministic control on community assembly peaks at intermediate levels of disturbance. Oikos 118, 471–479.

Malone, E.T., Abbott, B.W., Klaar, M.J., Kidd, C., Sebilo, M., Milner, A.M., et al., 2018. Decline in ecosystem $\delta^{13}C$ and mid-successional nitrogen loss in a two-century postglacial chronosequence. Ecosystems 21, 1659–1675.

Mann, D.H., 1983. The Geology of the Lituya Bay Glacial Refugium. PhD thesis University of Washington.

Mann, D.H., Gaglioti, B.V., 2024. The Northeast Pacific Ocean and Northwest Coast of North America within the global climate system, 29,000 to 11,700 years ago. Earth-Sci. Rev. 254.

Mann, D.H., Hamilton, T.D., 1995. Late pleistocene and holocene paleoenvironments of the North Pacific coast. Quat. Sci. Rev. 14, 449–471.

Mann, D.H., Streveler, G.P., 2008. Post-glacial relative sea level, isostasy, and glacial history in Icy Strait, Southeast Alaska, USA. Quaternary Res. 69, 201–216.

Mann, D.H., Ugolini, F.C., 1985. Holocene glacial history of the Lituya District southeast Alaska. Can. J. Earth Sci. 22, 913–928.

Milner, A.M., 2024. Glacier Bay National Park, Alaska: A natural laboratory to study stream development and succession following deglaciation. Adv. Ecol. Res. 70, 1–21. https://doi.org/10.1016/bs.aecr.2024.09.006.

Milner, A.M., 1987. Colonization and ecological development of new streams in Glacier Bay National Park, Alaska. Freshw. Biol. 18, 53–70.

Milner, A.M., 1994. Colonization and succession of invertebrates in a new stream in Glacier Bay, Alaska. Freshw. Biol. 32, 387–400.

Milner, A.M., 2016. conceptual model of community structure within glacier-fed rivers: 20 years on. In: Gilvea, D.T., Greenwood, M.T., Thoms, M.C., Wood, P.J. (Eds.), River Science: Research and Management for the 21st Century, pp. 156–170.

Milner, A.M., Bailey, R.G., 1989. Salmonid colonization of new streams in Glacier Bay National Park Alaska. *Aquaculture Fish. Manag.* 20, 179–192.

Milner, A.M., Fastie, C.L., Chapin, F.S., Engstrom, D.R., Sharman, L.C., 2007. Interactions and linkages among ecosystems during landscape evolution. Bioscience 57, 237–247.

Milner, A.M., Gloyne-Phillips, I.T., 2005. The role of riparian vegetation and woody debris in the development of macroinvertebrate assemblages in streams. River Res. Appl. 21, 403–420.

Milner, A.M., Knudsen, E.E., Soiseth, C., Robertson, A.L., Schell, D., Phillips, I.T., et al., 2000. Colonization and development of stream communities across a 200-year gradient in Glacier Bay National Park, Alaska, USA. Can. J. Fish. Aquat. Sci. 57, 2319–2335.

Milner, A.M., Magnusson, K., Veal, A.J., Brown L.E., 2024. Chironomidae (Diptera) community succession in streams across a 200 year gradient in Glacier Bay National Park, Alaska, USA. Adv. Ecol. Res. 70, 99–119. https://doi.org/10.1016/bs.aecr.2024.09.005

Milner, A.M., Petts, G.E., 1994. Glacial rivers: physical habitats and ecology. Freshw. Biol. 32, 295–307.

Milner, A.M., Picken, J.L., Klaar, M.J., Robertson, A.L., Clitherow, L.R., Eagle, L., et al., 2018. River ecosystem resilience to extreme flood events. Ecol. Evol. 8, 8354–8363.

Milner, A.M., Robertson, A.L., Brown, L.E., Sonderland, S.H., Mcdermott, M., Veal, A.J., 2011. Evolution of a stream ecosystem in recently deglaciated terrain. Ecology 92, 1924–1935.

Milner, A.M., Robertson, A.L., 2010. Colonization and succession of stream communities in Glacier Bay Alaska; What has it contributed to general ecological theory? River Res. Appl. 26, 26–35.

Milner, A.M., Robertson, A.L., Mcdermott, M.J., Klaar, M.J., Brown, L.E., 2013. Major flood disturbance alters river ecosystem evolution. Nat. Clim. Change 3, 137–141.

Milner, A.M., Robertson, A.L., Monaghan, K.A., Veal, A.J., Flory, E.A., 2008. Colonization and development of an Alaskan stream community over 28 years. Front. Ecol. Environ. 6, 413–419.

Milner, A.M., Sonderland, S.H., Brown, L.E., Clitherow, L., Eagle, L., Finn, D., et al. Looking forward; a synthesis of stream research undertaken in Glacier Bay. Adv. Ecol. Res. 71, next issue.

Molnia, B.F., 2007. Late nineteenth to early twenty-first century behavior of Alaskan glaciers as indicators of changing regional climate. Glob. Planet. Change 56, 23–56.

Monaghan, K.A., Milner, A.M. Disturbance and successional processes from spawning salmon and floods in Wolf Point Creek, Glacier Bay, Alaska. Adv. Ecol. Res. 71, next issue.

Monaghan, K.A., Milner, A.M., 2008a. Dispersal mechanisms of macroinvertebrates colonizing salmon flesh in a developing Alaskan stream. Acta Oecol.-Int. J. Ecol. 34, 65–73.

Monaghan, K.A., Milner, A.M., 2008b. Salmon carcasses as a marine-derived resource for benthic macroinvertebrates in a developing postglacial stream, Alaska. Can. J. Fish. Aquat. Sci. 65, 1342–1351.

Monaghan, K.A., Milner, A.M., 2009. Effect of anadromous salmon redd construction on macroinvertebrate communities in a recently formed stream in coastal Alaska. J. North. Am. Benthol. Soc. 28, 153–166.

Monteith, D., 2017. Understanding landscape change using oral histories and tlingit place-names. Critical Norths: Space, Nature, Theory. pp. 269–285.

Pickett, S.T.A., 1982. Population-patterns through 20 years of old field succession. Vegetatio 49, 45–59.

Robertson, A.L., Milner, A.M., 1999. Meiobenthic arthropod communities in new streams in Glacier Bay National Park, Alaska. Hydrobiologia 397, 197–209.

Robertson, A.L., Milner, A.M., 2001. Coarse particulate organic matter: a habitat or food resource for the meiofaunal community of a recently formed stream? Arch. Fur Hydrobiol. 152, 529–541.

Robertson, A.L., Milner, A.M., 2006. The influence of stream age and environmental variables in structuring meiofaunal assemblages in recently deglaciated streams. Limnol. Oceanogr. 51, 1454–1465.

Rumore, G. 2009. A natural laboratory, a national monument: carving out a place for science in Glacier Bay, Alaska, 1879–1959. PhD, University of Minnesota.

Sonderland, S.H., Milner, A.M., 2024. Salmon lice (*Lepeophtheirus salmonis*) as a food source for juvenile salmonids in Glacier Bay, southeast Alaska. Adv. Ecol. Res. 70, 121–136. https://doi.org/10.1016/bs.aecr.2024.09.001.

Sonderland, S.H., Milner, A.M., 2024. The spatiotemporal dynamics of the physical habitat template at a watershed scale chronosequence in Glacier Bay, southeast Alaska. Adv. Ecol. Res. 70, 23–66. https://doi.org/10.1016/bs.aecr.2024.09.002.

Walker, L.R., Wardle, D.A., Bardgett, R.D., Clarkson, B.D., 2010. The use of chronosequences in studies of ecological succession and soil development. J. Ecol. 98, 725–736.

Wiles, G.C., Lawson, D.E., Lyon, E., Wiesenberg, N., D'arrigo, R.D., 2011. Tree-ring dates on two pre-Little Ice Age advances in Glacier Bay National Park and Preserve, Alaska, USA. Quat. Res. 76, 190–195.

Windsor, F.M., Wilkes, M., Brown, L.E., Robertson, A., Milner, A.M. Invertebrate functional trait diversity along successional gradients in Glacier Bay. Adv. Ecol. Res. 71, next issue.

CHAPTER TWO

The spatiotemporal dynamics of the physical habitat template at a watershed scale chronosequence in Glacier Bay, southeast Alaska

Svein Harald Sønderland[a],* and Alexander M. Milner[a,b]

[a]School of Geography, Earth and Environmental Sciences, University of Birmingham, Edgbaston, Birmingham, United Kingdom
[b]Institute of Arctic Biology, University of Alaska Fairbanks, Fairbanks, AK, United States
*Corresponding author. e-mail address: snarrald@gmail.com

Contents

1. Introduction	24
1.1 Landscape ecology	24
1.2 Remote sensing	25
1.3 Scale and resolution	27
1.4 Physical habitat template (PHT,t) of a riverscape	29
2. Methods	30
2.1 Study area	30
2.2 DTM mapping and data analysis	30
2.3 Statistics	35
3. Results	35
4. Discussion	42
4.1 Eco-geomorphology	42
4.2 Framework for a remote sensed PHT at a point in time	49
4.3 Directional development of the PHT,t	51
5. Summary	56
Acknowledgements	57
Appendix 1. Watershed variables correlation table	58
References	60

Abstract

Spatial and temporal processes within watersheds need to be characterised to understand river system development. In Glacier Bay, southeast Alaska, following glacial retreat, basic physical morphological features of five watersheds of different ages were analysed using interferometric synthetic aperture radar (IFSAR) and geographic information systems (GIS). The initial conditions after glacial retreat, especially the location of the watershed in the landscape, determined the temporal effect and successional processes, with the initial topology as the main driver in watershed development. Median watershed elevation and mean watershed slope significantly influenced drainage density negatively and had also a negative impact on vegetation cover, and total waterbodies influenced negatively the total number of streams. Vegetation cover influenced positively drainage density, and stream frequency within watersheds, due to the development of topsoil caused by litter fall. Watershed age influenced negatively relief ratio while positively influenced total number of streams significantly. Total number of streams was therefore the best indicator of watershed development in Glacier Bay. Watershed development within Glacier Bay is a result of nonlinear deterministic chaos. The mosaic of physical habitat variation in the watershed is connected by a stream network creating heterogeneous habitats which influences the biotic community. Here we present a conceptual framework model of a four dimensional physical habitat template at a point in time (PHT,t), and its directional development. The conceptual framework model helps us understand the interconnectedness within a stream network and its watershed, providing a more holistic view of connectivity and landscape processes in a space-time continuum and thus contributes to watershed management and monitoring.

1. Introduction
1.1 Landscape ecology

There are many connections across terrestrial and aquatic systems, but ecologists have historically viewed these ecosystems separately (Soininen et al., 2015). Links between physical habitat and ecological responses in rivers are widely recognised, but are typically poorly quantified (Orr et al., 2008). Rapid glacial recession in Glacier Bay creates a chronosequence with a unique opportunity to study landscape evolution and watershed development. Succession and colonisation of vegetation and invertebrates both inside and outside the stream network can therefore be related to the post-glacial processes, providing a unique opportunity to study these processes on a larger spatial scale within a shorter timeframe (Kling, 2000). Morphological features of watersheds are important environmental variables, influencing both colonisation and succession in streams and associated ecosystems, and thus understanding how processes are affected by landforms yield some power in predicting ecosystem behaviour (Swanson et al., 1988). Analysing and quantifying surface morphology in terms of landform characteristics are essential for understanding physical, chemical, and biological processes occurring within the landscape (Blaszczynski, 1997), since the interplay of physical and biological factors develop natural ecosystems (Swanson, 1980). Topography of watersheds has a major impact on the hydrological, geomorphological and biological processes operating within the landscape (Moore et al.,1991). Horton (1932) proposed five descriptive factors of a watershed related to its hydrology; (1) morphologic-, (2) soil-, (3) geologic-structural-, (4) vegetational- and (5) climatic-hydrologic factors, since hydromorphological turnover processes is recognised as key for development of distinct habitat patterns (Hohensinner et al., 2011).

Conditions of the initial site in Glacier Bay influence the rate of change and the final state of community composition and productivity (Chapin et al., 1994). Macroinvertebrate community assemblage in newly emergent Glacier Bay streams has been found to be initially be strongly deterministic, while microcrustacean assemblies were more stochastic (Milner et al., 2011). River channel morphology is dependent on the geology and climatic environment over long periods of time, but during shorter periods channel morphology is an independent variable influencing river channel hydraulics (Schumm and Lichty, 1965). Different effects of disturbance on a given biological response variable, requires appropriate spatial and temporal scales to be detected (Poff and Ward, 1990). These scales create a habitat that provides a template where characteristic species traits of macroinvertebrates are forged by evolution (Townsend and Hildrew, 1994). Factors operating at several spatial and temporal scales influence the physical habitat (Medeiros et al., 2008), and the heterogeneity of the environment is regarded as one of the most important factors governing community structure (Poff and Ward, 1990; Stein et al., 2014; Yang et al., 2015). Phillips (2007) provided the possibility of multiple outcomes as a "perfect landscape", because cause and effect in developing landforms is a function of time and space, as either dependent or independent of variables as time and space change (Schumm and Lichty, 1965). Therefore, natural landscapes will typically never achieve equilibrium since processes may change faster than the landscapes can respond (Montgomery and Dietrich, 1992).

1.2 Remote sensing

Remote sensing is a powerful method to characterise abiotic and biotic variables that change over time (Rocchini et al., 2013). Developing spatially continuous information across hierarchical scales by employing remote sensing and geospatial tools can help understand riverscape reaches in their broader network context (Glassic et al., 2024). Modern geomorphometry focuses on extraction of land surface variables and objects/features from digital topography (Wilson, 2011). One of the core products of photogrammetry and remote sensing is Digital Elevation Models (DEM's) (Schindler et al., 2011). Estimation base data for a DEM are terrain points, historically acquired manually in the field, but nowadays technologies for large area 3D points exist (Schindler et al., 2011). Digital representation of surface topology is useful to describe morphological features, and as deviation between elevation at one specified point and its true value diminishes, accuracy increases. Different algorithms within software have been developed to analyse DEM's and to facilitate incorporation of field measurements. The technologies and methods used to collect the source data, the preprocessing algorithms applied, and the land surface characteristics itself will determine the frequency and magnitude of errors (Wilson, 2011).

Drainage networks are significant in modelling landscape development and drainage basin hydrology (Mark, 1984), and GIS can be used to study landscape connectivity (Marston, 2010), where stream networks are essential in linking watersheds and larger landscapes together. DEMs are therefore useful data sources from which drainage characteristics can be derived automatically with the help of software tools (Jenson, 1985; Jenson and Domingue, 1988; Tarboton and Ames, 2001). Riverscape approaches presents logistical and technical challenges depending on the spatiotemporal scope of the study (Torgersen et al., 2022). This is due to the potentially large data acquisition with higher resolution, or/and the time spent in the field including logistical problems in remote areas. A variety of remote sensing instruments/sensors are available for studying riverscapes (Mertes, 2002), and the characteristics of the instrument/remote sensors carrying platform (satellite, airborne, UAS and mobile/static sensors) play a significant role in how efficiently the object space can be observed (Toth and Jóźków, 2015). Resolution of the point cloud/volume at a certain scale will determine what processes and patterns that can be detected, and field/aerial observation are important to ensure that the delineation is correct. Field mapping is the most accurate way to determine drainage networks, despite its impracticality for large areas (Tarboton and Ames, 2001).

Riverscape examination changes as technical instruments expand, from microprobes to satellites, to be able to examine scales from microhabitats, to channel units, to the valley watersheds, spatial and temporal relationships among biota, hydrology and geomorphology (Mertes, 2002). To characterise changes in space and time that act on the Earth's system, continuous geodetic observations are fundamental (Altamimi and Gross, 2017). Correct georeferencing is essential to measure accurately the riverscape horizontal and vertical position and its variation through time. Differences in georeferencing between two independent DEMs are more a norm than an exception, both in plain planimetry and height (Schindler et al., 2011).

Better georeferencing performance is needed with more sophisticated remote sensing sensors (Toth and Jóźków, 2015) and will be able to reconstruct the measured object or riverscape more accurately. Technologies and platforms within remote sensing have evolved rapidly (Fassnacht et al., 2024) and will evolve further and increase resolution and capabilities to better measure spatial and temporal data.

1.3 Scale and resolution

Ferrari and Ferrarini (2008) concluded that ecosystem and landscape ecology differ from a theoretical viewpoint, due to the difference in spatial and temporal scales used. Ecosystems of rivers are therefore more and more viewed as dynamic riverscapes (Fausch et al., 2002; Malard et al., 2006, 2000; Ward et al., 2002), and being able to measuring variability across multiple scales can take the riverscape concept from theory to reality (Carbonneau et al., 2011). Changing the scale of a variable, changes the variance of that variable, scale and resolution are therefore two main factors controlling how and if processes and patterns are detectable (Wiens, 1989). When predicting interactions between physical and biological processes the determination of scales is therefore essential (Reinhardt et al., 2010). Frissell et al. (1986) developed a hierarchal classification system to view stream habitats in the context of the watershed. Spatial scales within the watershed with increasingly accurate remote sensing, contribute to connect Frissel's classification system together within a continuous spatiotemporal watershed and their heterogeneous habitats. Digital terrain model (DTM) raster is here shown as a two-dimensional cell volume system in a modified Frissell et al. (1986) hierarchal classification system (Fig. 1) with the same resolution at different scales, visualising how this controls the spatial and temporal data. Scaling is therefore more a question of the quality of the remote sensing data, as this will determine the scale of what can be observed. Remote sensing can be used in relation to most scales, but to cover larger spatial scales like a watershed, use of satellites and aircraft remote sensing are a more reasonable method, as handheld remote sensing is very time consuming. The influences from surrounding factors are almost infinite and therefore difficult to observe, and the importance of scaling will be related to the physical and biological effects induced on the habitats. These effects from morphology and geology can originate from around the perimeter of the watershed and would therefore not be accounted for on a reach scale (Fausch et al., 2002; Talluto et al., 2024; Wiens, 2002), and mapping on a

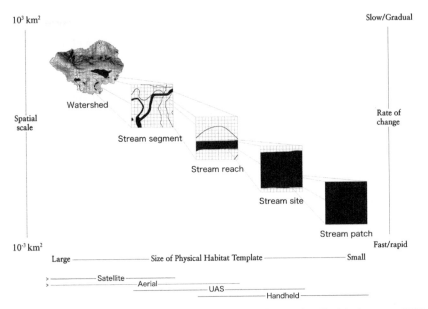

Fig. 1 Functional classification of a river system based on scale, all with the same DTM raster cell system resolution. *After Frissell, C., Liss, W., Warren, C., Hurley, M., 1986. A hierarchical framework for stream habitat classification: viewing streams in a watershed context. Environ. Manag. 10, 199–214. https://doi.org/10.1007/BF01867358; Maddock, I., 1999. The importance of physical habitat assessment for evaluating river health. Freshw. Biol. 41, 373–391. https://doi.org/10.1046/j.1365-2427.1999.00437.x.*

watershed scale, biota and nutrient transfer can more readily and more accurately be assessed, without extrapolation from reach sections.

The key factor in the evolution of mountain landforms is channel initiation (Imaizumi et al., 2010), and drainage network extraction depends primarily on the accuracy of the maps used and the identification of the channel initiation (Lin et al., 2006). The closer a channel commences to the drainage divide, the more finely dissected the watershed will be, as greater number of channels will occupy a unit area (Montgomery and Dietrich, 1988). Automated channel initiation is therefore manly limited by the threshold of the contributing area (O'Callaghan and Mark, 1984). The stream functions like "blood vessels" into the landscapes, bringing nutrients downstream and into to the ocean, while making a conduit for fish, insects and animals upstream. Nutrients, materials and biota can also use rain, wind and other biological conduits for looping upstream. All these interactions create a mosaic of habitats within the landscape. To detect the effects of disturbance on a given biological response variable the appropriate spatial and temporal scale is required (Poff and Ward, 1990). Advances in remote sensing and field equipment implementation of new technologies can now more readily address both scale and resolution, and smaller changes within the watershed can be characterised, such as vegetational change, geomorphology and other factors only before described on a reach scale.

1.4 Physical habitat template (PHT,t) of a riverscape

Poff and Ward (1990) placed the stream (drainage network) in a watershed context, since its characteristics reflect the geology and climatic history of the watershed and associated bio-geoclimatic region. Riverscapes are dynamic environments (Glassic et al., 2024; Malard et al., 2006; Wiens, 2002), and ecosystem system characteristics are not the same at relatively fine scales as they are relatively broad scales (Wiens, 1989). Terrestrial and aquatic complexities within the watershed are integrated (Milner et al., 2007), and seamlessly map both domains are therefore required to fully understand these systems (McKean et al., 2008). The living space of the stream biota is the physical habitat, a spatially and temporally dynamic entity (Maddock, 1999). Many concepts has been developed attempting to describe the interactions and energy flow between heterogenous habitats throughout the riverscape (Junk et al., 1989; Stanford and Ward, 1993; Vannote et al., 1980; Ward and Stanford, 1983; Webster and Patten, 1979). Streams are strongly hierarchical and patchy systems (Poole, 2002), and to investigate patterns and processes Pringle et al. (1988) presented a framework that considers how patch specific characteristics determine biotic and abiotic processes over various scales. Romme and Despain (1989) found that patchiness, combined with differential rates of succession maintained a considerable level of heterogeneity. The continuum of the biosphere (Rull, 2014) gives that everything is interconnected through time, and Southwood (1977) presented the habitat as a template for ecological strategies, consisting of a continuum of habitat heterogeneity, using space and time as two basic dimensions. Ward (1989) conceptualised the dynamical and hierarchical nature of lotic ecosystems into a four-dimensional framework, consisting of longitudinal, lateral, vertical and time, since lotic ecosystems are open systems and in effect highly interactive with their surroundings. To narrow the gap between theory and practice, system-specific realism needs to be incorporated in otherwise predominantly conceptual studies, in addition to studying environmental change scenarios (Schiesari et al., 2019).

The aim of this paper is to present a conceptual framework model of a PHT,t, and its direction of development based on the morphological features within watersheds found in Glacier Bay. The conceptual framework model of a PHT,t was here created based on remote sensing and GIS and will help provide insight into spatial and temporal changes in physical habitats. Minkowski space (Minkowski, 1908) was used to describe an integration of a four-dimensional point/resolution layer system (x,y,z,t) that will create a three dimensional structural physical habitat template at a point in time. Combining field and data to a continuum system will provide better understanding of the processes and patterns on a spatial-temporal scale. The conceptual framework model describes the development with time, and how space and time is related to scale and resolution and explains the directional development in a four-dimensional space, since the temporal component is stored in space-time geometry (Hazelton et al., 1990).

2. Methods
2.1 Study area

Glacier Bay has a strong tidal activity and a maritime climate, situated in the Pacific temperate rainforest region, with mild winters and cool summers. Glacier Bay consists of a complex fjord system with two major arms: (1) the West arm approximately 59 km in length and (2) the East arm 48 km. Study streams were located from the upper part of the east arm to the lower bay area. The studied watersheds were Stonefly Creek (SFC), Wolf Point Creek (WPC), Ice Valley Stream (IVS), Berg Bay South Stream (BBS), and Rush Point Creek (RPC) (Fig. 2) ranging in age from 46 to 213 y in 2024 (Table 1 age is from 2012 when the IFSAR data were collected). Stream age was determined using satellite and aerial photos, historical data, journal articles and unpublished data outlined in Milner et al. (2000).

2.2 DTM mapping and data analysis

Digital elevation models (DEM's) were used to automatically map watersheds and stream network in Glacier Bay. IFSAR 5-m DEM's from Fugro EarthData, Inc, were used to calculate morphological data. ERDAS IMAGINE v. 15.1 was used to mosaic the DEM's together, before areas with the watersheds of interest were cut out to limit the region size. Various variables were then calculated to produce valuable information regarding watershed's geomorphological features. Watersheds were then extracted in GRASS 7 (Geographic Resources Analysis Support System) with the modules *r.stream.basins* and *r.water.outlet*, after using *r.stream* extract

Spatiotemporal dynamics

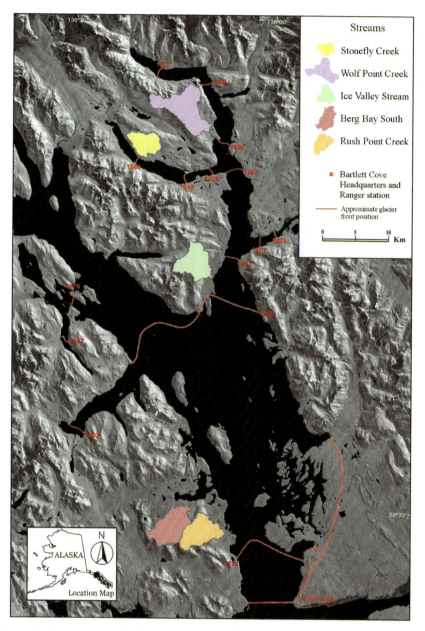

Fig. 2 Study watersheds in Glacier Bay, Alaska. Map lines delineate study areas and do not necessarily depict accepted national boundaries.

Table 1 Derived characteristics of the five study watersheds (Stream age based on the year the IFSAR data was collected (2012) and equates to when the steam mouth was uncovered).

Variables	Watersheds					PCA NR	Reference	Formula
	SFC	WPC	IVS	BBS	RPC			
Age (Stream)	34	60	136	176	201	N1	-	-
Area (km^2) (A)	13.11	29.68	19.04	22.44	20.14	N2	Grass	Attribute Table Manager
Area elevation bellow 50m (km^2)	3.37	2.86	1.05	2.12	4.17	N3	Grass	r.mapcalc
Channel storage ratio (ρ)	0.55	0.60	1.01	1.14	0.72	N4	Horton (1945)	$\rho = Rl/Rb$
Compactness	1.54	2.00	1.68	1.81	1.65	N5	Grass	Attribute Table Manager
Constant of channel maintenance (Ccm)	0.60	0.85	0.70	0.65	0.62	N6	Schumm (1956)	$Ccm = 1/Dd$
Drainage density (Dd)	1.66	1.17	1.42	1.53	1.61	N7	Grass	r.stream.stats
Elongation ratio (Re)	0.64	0.46	0.48	0.51	0.48	N8	Schumm (1956)	$Er = (2\sqrt{A/\pi})/Lw$
Fineness ratio (Rf)	1.10	0.90	1.03	1.13	1.24	N9	Melton (1957)	$Rf = Ls/P$
Form factor ratio (Ff)	0.32	0.17	0.18	0.21	0.18	N10	Horton (1932)	$Ff = A/Lw^2$
Hack main channel (km)	6.14	11.84	9.47	9.55	9.73	N11	Grass	r.stream.stats
Length of overland flow (Lo)	0.303	0.429	0.353	0.329	0.311	N12	Horton (1945)	$Lo = 1/2Dd$
Main channel azimuth	203.81	95.18	85.93	21.02	58.19	N13	Grass	Attribute Table Manager
Main channel sinuosity	1.67	1.34	1.74	1.69	1.56	N14	Grass	Attribute Table Manager
Max watershed slope in degrees	63.90	73.63	74.24	69.85	69.06	N15	Grass	r.univar
Mean Bifurcation ratio (Rb)	4.80	5.38	3.17	3.17	3.59	N16	Grass	r.stream.stats

Metric						Code		Reference	Tool
Mean Length ratio (Rl)	2.66	3.20	3.18	3.62	2.60	N17	Grass		r.stream.stats
Mean watershed elevation (m)	214.08	357.25	314.28	228.82	232.80	N18	Grass		r.univar
Mean watershed slope in degrees	16.09	21.78	20.84	18.08	17.45	N19	Grass		r.univar
Median watershed elevation (m)	94.18	291.71	271.47	203.58	188.86	N20	Grass		r.univar
Median watershed slope degrees	13.63	20.68	20.81	16.44	15.11	N21	Grass		r.univar
Melton ruggedness number (RR)	0.195	0.227	0.217	0.156	0.173	N22	Melton (1965)	$RR = H/\sqrt{A}$	
Perimeter (km) (P)	19.79	38.63	26.06	30.33	26.17	N23	Grass		Attribute Table Manager
Relief ratio (Rr)	0.11	0.093	0.093	0.071	0.074	N24	Schumm (1954)	$Rr = H/Lw$	
Stream frequency (Fs)	2.22	1.08	2.01	1.70	2.43	N25	Grass		r.stream.stats
Stream order	3	3	4	4	4	N26	Grass		r.stream.stats
Total Length of first order streams (km) (Ls1)	12.24	17.70	15.28	14.26	17.39	N27	Grass		r.stream.stats
Total length of streams (km) (Ls)	21.67	34.61	26.94	34.12	32.40	N28	Grass		r.stream.stats
Total number of streams (Ns)	29	32	38	38	49	N29	Grass		r.stream.stats
Vegetation in the watershed (%)	72.43	50.62	70.23	71.97	77.21	N30	Grass		r.mapcalc
Waterbodies (km²)	0.74	1.48	0.01	0.10	0.03	N31	Grass		r.mapcalc
Waterbodies in the watershed (%)	5.63	4.99	0.02	0.44	0.13	N32	Grass		r.mapcalc
Watershed length (km) (Lw)	6.42	13.25	10.20	10.42	10.48	N33	Grass		Attribute Table Manager
Watershed relief (m) (H)	705.91	1235.13	947.96	738.27	774.45	N34	Grass		r.univar

to find the stream mouth, and high tide from field observations were used as outlet points for the streams. However, only BBS and RPC would get a larger watershed area with low tide, due to streams flowing together between low and high tide. Attribute Table Manager in GRASS 7 was used to calculate area, perimeter and compactness from the polygon created from the raster map (Table 1).

Stream delineation was undertaken with *r.stream.extract*, which used multiple flow directions (MDF) to give a more accurate stream network (Jasiewicz and Metz, 2011). After first stream extraction from the DTM, errors found in the newly created stream network where changed based on field and aerial observations and patched in the DTM with *d.rast.edit*, before delineated again with *r.stream.extract* in GRASS 7. The only stream where most of the tributaries were known was SFC, and by georeferencing a map created in ArcMap in 2004, a more accurate threshold would be around 0.5%. This will vary from stream to stream, and thus 1% flow accumulation was used to avoid possible overestimation. This gives a strong underestimate of the stream network, but will be the same for all watersheds, and make comparison possible. Stream network calculations were undertaken with *r.stream.stats*, and elevation and slope (*r.slope.aspect*) data from the raster maps (DTMs) was calculated with *r.univar*, and contour line map was done with *r.contour*. While main channel azimuth and sinuosity and watershed length was calculated with Attribute Table Manager after polyline was created based on main channel polyline from Hack in *r.stream.stats*.

Vegetation area was calculated from the watershed polygon, derived from digital surface model DSM minus DTM with *r.mapcalc*, and provide a good indication on vegetational coverage. Due to not having the x-band to correct for backscatter, there were smaller errors in the height of the vegetation, and this was therefore not used as a variable for vegetation succession. Vegetation cover was likely a minor overestimate since we cannot correct for backscatter. The total amount of lentic waterbodies (lakes, ponds including kettles) was calculated from polygons of known waterbodies and thus is likely an underestimate. Unknown smaller waterbodies might have been missed, since we could not correct for backscatter (x-band missing) in areas with high elevation differences and dense vegetation.

Topographic wetness index (TWI), was calculated with *r.topidx* in GRASS 7 by $\ln(a/\tan\beta)$ where a is the upslope area draining through a certain point per unit contour length, and $\tan\beta$ the local topographic surface slope (Δ vertical /Δ horizontal). Many derived quantities have been developed, and the few used herein were calculated based on the watershed characteristics found with GRASS 7, with their formulas displayed in Table 1. These are often related to relief, form and texture of the watershed. Relative relief (Rr) expressed as the height (H) of the watershed divided by its length (Lw) (Schumm, 1954). Melton (1965) characterised the ruggedness (RR) of a watershed as the height between the maximum and minimum elevation (H) of the watershed on the squared area (A). Fineness ratio (Rf) is the total lengths of steams (Ls) to the watershed perimeter (P) and is a dimensionless measure of fineness of topographic texture relative to watershed size (Melton, 1957). Form factor (Ff) is the ratio of the drainage area (A) to double the watershed length (Lw^2) (Horton, 1932). Elongation ratio (Er) is the ratio between the diameter of circle with the same area (A) as the watershed on the watershed length (Lw) (Schumm, 1956). Length of overland flow (Lo) is half of the drainage density (Dd) and is the distance water travel over the ground surface before reaching definite stream channels (Horton, 1932). The Rho ratio (ρ) is the mean length ratio (Rl) on the mean bifurcation ratio (Rb) and a measure of channel storage (Horton, 1945). Constant of channel maintenance (Ccm) and is the minimum limiting area (1 m) required for drainage channel development (Dd) (Schumm, 1956).

2.3 Statistics

Spearman-Rank correlations were calculated for stream variables (Appendix 1) and principal component analysis (PCA) was used to help assess directional relationship between the different watersheds using (sjPlot 2.8.16) package (Lüdecke, 2024) and (factoextra 1.0.7) package (Kassambara and Mundt, 2020) in R Statistical software (2024.4.1.748) (R Core Team, 2024).

3. Results

The study watersheds were markedly different from each other according to their spatial location following deglaciation. SFC (13.1 km^2) sits on the margin of the U-shaped valley creating Wachusett Inlet, with an upper clearwater lake (0.52 km^2) and a lower lake (0.09 km^2) still influenced by glacial runoff from remnant ice. WPC (29.7 km^2) was created by a side glacier, next to the main fjord, with a paternoster lake system at the head of the watershed. One larger main lake (1.20 km^2) and two smaller lakes above (0.13 km² and 0.09 km²) feed the stream. IVS (19.0 km²) is located on the side of Muir Inlet and has a main channel with an even slope up to the east side of the watershed and was the watershed least influenced by lakes with presently only a few wetlands and smaller muskeg waterbodies. BBS (22.4 km²) is a watershed created from the retreat of a major glacial outflow which flowed east into the main fjord, and has several glacial erosion cirque formations, and a cirque kettle lake (0.09 km²) at the upper end of the watershed. Like BBS, the RPC (20.1 km²) watershed was created after the retreat of a side glacier flowing eastwards into the main fjord with a few glacial erosion cirque formations, and a hanging valley at the west end of the watershed. The RPC watershed supported smaller waterbodies due to several beaver dams and two small lakes. All streams have barriers in the mid-lower main channel except for IVS. Vegetation cover was different from watershed to watershed, with taller more mature vegetation in older watersheds. A summary of watersheds characteristics that were determined is in Table 1 and discussed below.

The two youngest watersheds (SFC and WPC) were stream order 3, while the rest of the watersheds were order 4 (Fig. 3). Vegetation cover (Fig. 4) occur in areas with lower slope in degrees (Fig. 5). SFC and RPC had the highest low topographic wetness index (TWI) values (Fig. 6) and the lowest mean slope and the highest drainage densities and was the watersheds with large areas at < 50 m elevation, which contained larger parts of the main channel. SFC and WPC had most waterbodies within the watersheds and can be observed within the 25 m contour lines (Fig. 7). WPC and IVS have areas within the watershed with the lowest and the highest TWI and are the streams with the highest mean watershed slope and the lowest drainage density (<1.5) and are the watersheds with highest relief and ruggedness. Both these streams also had the lowest percent of vegetation cover. All five streams had a mean bifurcation ratio > 3, with WPC exceeding 5. BBS showed the highest channel storage ratio, lowest relief ratio and ruggedness number, while SFC and WPC had the lowest channel storage ratio and both systems has larger lakes.

Watershed age positively correlated with total number of streams in the watershed significantly and positively ($p < 0.1$) with stream order and negatively ($p < 0.1$) with relief ratio and main channel azimuth. Percentage vegetational cover correlates positively ($p < 0,1$) with drainage density, stream frequency and fineness ratio, while negatively correlating with constant of channel maintenance and length of overland flow, and with median elevation and mean elevation slope, and otherwise negatively with all factors

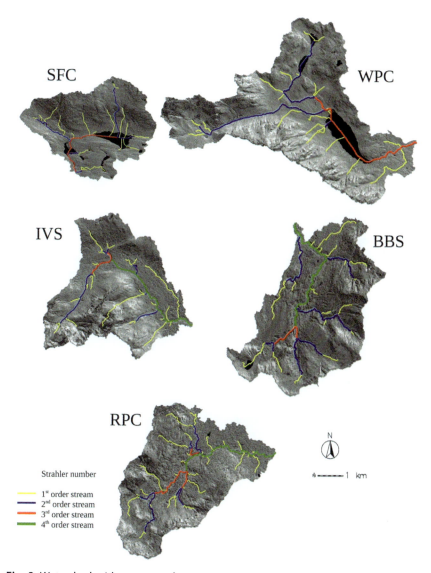

Fig. 3 Watershed with stream order.

relating to elevation. Percent water bodies correlated negatively (p < 0.1) with stream order and total number of streams. Water bodies area (km^2) positively correlated (p < 0.1) with mean bifurcation ratio and negatively with stream order. Total length of first order streams negatively correlated (p < 0.01) with watershed elongation ratio and form factor significantly and positively correlated (p < 0.1) with mean watershed elevation, Hack main

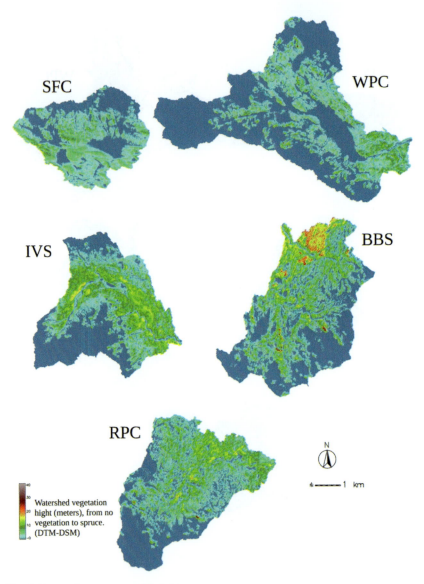

Fig. 4 Vegetation cover and height.

channel, watershed length and relief. Median watershed elevation and mean watershed slope negatively correlated (p < 0.05) with drainage density significantly and negatively (p < 0.1) with elongation ratio and form factor. Main channel azimuth positively correlated (p < 0.01) with relief ratio significantly.

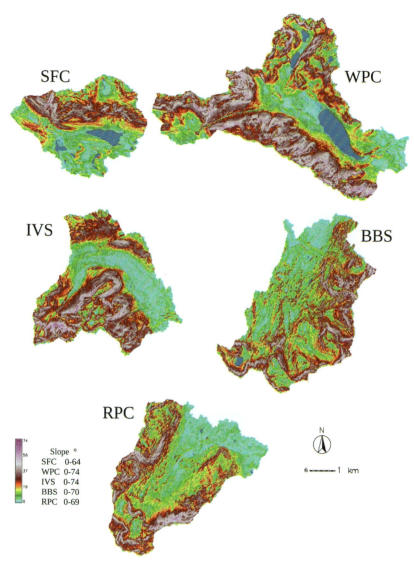

Fig. 5 Watershed slope in degrees.

Main channel longitudinal profiles for each stream show the elevation (m) and length (m) of the main channel (Fig. 8), and barriers along the main channel. The length of the main channel (Hack) of WPC was the longest registering 11.84 km, and SFC the shortest at 6.14 km. There are large differences in the relief ratio between the different streams, where the youngest stream had the highest and the two oldest having the lowest relief ratio values

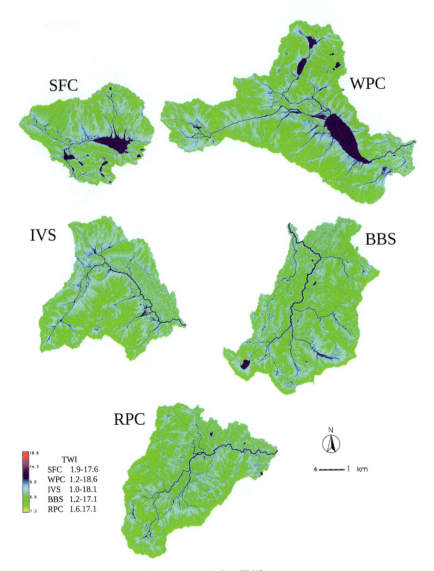

Fig. 6 Watershed topographic wetness index (TWI).

(Table 1). Correlations among the watersheds are in Appendix 1. The PCA space (Fig. 9) shows the position of the watersheds in relation to each other and show that BBS and RPC are the most similar watersheds, while SFC and WPC are positioned furthest from each other in PCA space. The first two components explain 53% and 30% of the variance (total 83%). Contributing variables to PC1 and PC2 (Fig. 10A and B) above the line of relevance (dotted

Spatiotemporal dynamics

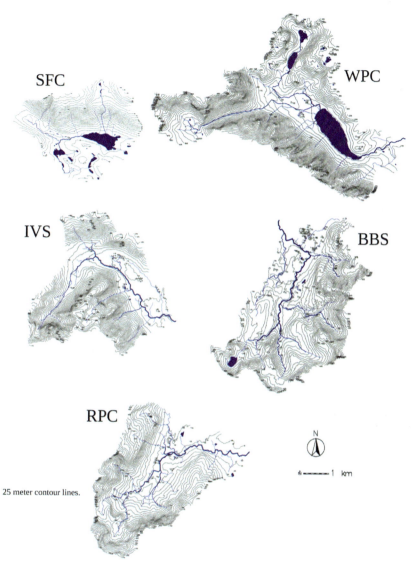

Fig. 7 Watersheds with 25 m contour lines, lakes, ponds, water bodies and the stream system.

line assuming uniform distribution) show that drainage density contributed the most to PC1, which describe the total length of streams on the area of the basin. The next variables relate to drainage density (e.g. constant of channel maintenance and length of overland flow) and then watershed relief and perimeter. Elevation variables come next before vegetation and the max

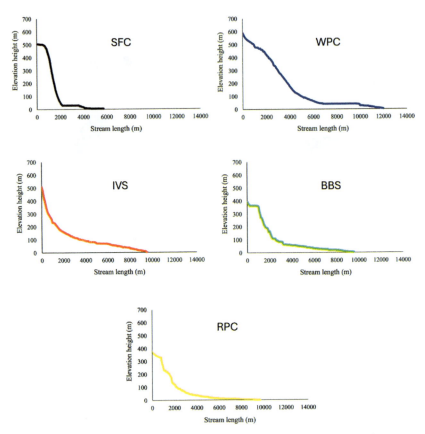

Fig. 8 Main channel profile (Distance x axis and elevation y axis are in m).

watershed slope before lastly variables describing the form of the watershed. The most important variable contribution to PC2 was stream age and then the waterbodies, stream order, azimuth, mean bifurcation ratio, total number of streams and channel storage ratio, then less important contributions from form factor, elongation ratio and ruggedness number.

4. Discussion

4.1 Eco-geomorphology

Morphometry of landforms is related to causative factors of mantle characteristics, climate, lithology and vegetation density (Melton, 1957), and watershed forms change with time during geomorphic development

Spatiotemporal dynamics

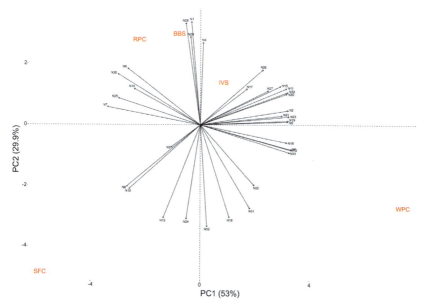

Fig. 9 PCA biplot of studied watersheds in Glacier Bay and their variables (description of N1-N34 are located in Table 1).

(Schumm, 1956). Many different forms/shape related factors and ratios has been developed to describe watersheds, since overall watershed drainage shape is geometrically related to the patterns of the drainage network, which is determined by many factors including relief, geology, climate and vegetation (Hack, 1957). Form factor indicates the flood regime of the watershed stream, however, for watersheds of irregular form and especially permeable soils form factor will not be a sensitive indicator of hydrologic characteristics (Horton, 1932). The elongation ratio describe the roundness of the watershed and as the ratio approaches 1 the shape approaches a circle (Schumm, 1956). GRASS uses Gravelius compact coefficient to calculate compactness, where 1 is perfectly round and then values increase with irregular and elongated watersheds (Sassolas-Serrayet et al., 2018). WPC had the lowest form factor and elongation ratio whereas SFC had the highest and reversed for compactness. SFC was the roundest of the studied watersheds; a rounder watershed will drain quicker than an elongated one, and the runoff from different location will arrive at the mouth at the same time resulting in greater peak flow (Schroeder et al., 2015).

Topographic texture and average slope influencing factors can be considered as fundamental in fluvial landscape control (Melton, 1957), and stream

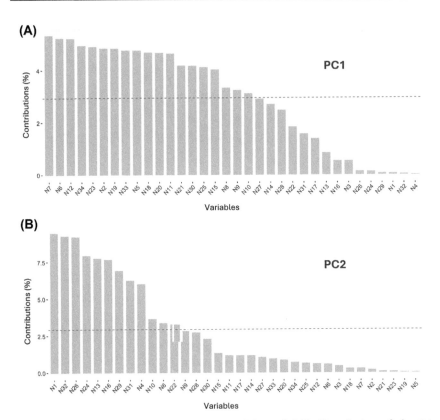

Fig. 10 (A, B) The variables contribution to PC1 and PC2 (Description of the 34 variables are found in Table 1).

profiles are well adjusted to carry away watershed erosion products at rates determined by the initial relief, time and geology of the watershed (Hack, 1957). Ruggedness number (relative relief) can vary from 0 to a large number, but generally third or fourth order watersheds rarely exceeds one (Melton, 1965). WPC and IVS had higher ruggedness number than SFC, and BBS had the lowest of the studied watersheds. The relief ratio seems a better measure for estimating the denudation and glacial isostatic adjustments with time, even though the maximum timeframe in Glacier Bay is only a few hundred years. Higher relief ratio in SFC, WPC and IVS could mean higher sediment loss, especially in SFC, since relief ratio is related to valley-side slopes, stream gradients and drainage density, which all are known to influence sediment loss (Schumm, 1954). Constant of channel maintenance decreases with increasing erodibility (surface erosion) (Schumm, 1956) with SFC having the lowest value. Ratio of measuring fineness of topographic texture was highest in RPC and BBS, and can be seen in PCA to be closely correlated to percent vegetation while negatively correlated with ruggedness.

Watershed topology influences the hydrologic cycle, particular runoff, and the profile of the watershed drainage provide the best basis to compare elevation with hydraulic characteristics of the watershed (Horton, 1932). Drainage density is the total length of streams per unit area and an excellent indicator of the surface permeability of the watershed (Horton, 1932) and also describes the channel dissection of the watershed well (Mutzner et al., 2016). Ranging from 1.5 to 2.0 for steep, impervious watersheds with high precipitation, down to 0 or close to 0 for watersheds with permeable surfaces where precipitation ordinarily is taken into the soil through infiltration (Horton, 1932), and rarely exceeds 3 (Horton, 1945). Higher drainage density allows the landscape to drain quicker and causes larger peak flows (Schroeder et al., 2015). Drainage density was lowest in WPS and IVS, the watersheds having the highest length of overland flow, suggesting higher infiltration rates and with less direct surface runoff (Horton, 1932). Length of overland flow is hydrologically of great importance, especially to flood intensity from small areas (Horton, 1932). However the greater the slope the lower the infiltration rate, since gravity pulls more across the surface and less into the surface, and again the steeper the slope the quicker the response and higher peak discharge (Schroeder et al., 2015). WPS and IVS watersheds had higher relief and ruggedness and length of overland flow, both watersheds had larger amounts of observed sediments. IVS had the lowest drainage density with variable flow and highly permeable stream bed, as water was absent from the stream bed close to the mouth on several occasions. TWI is commonly used to quantify the control of topology on hydrological processes (Sørensen et al., 2006). Moore et al. (1993) found the terrain attributes, slope and wetness index, to correlate the most with surface soil attributes. Larger low slope angle areas prone to water accumulation are linked to high TWI, while steep slope well drained areas prone to be dry are linked to low TWI (Mattivi et al., 2019). Higher TWI value regions have therefore higher probability for flooding (Khosravi et al., 2019). WPC and IVS have the lowest and highest TWI and can be expected to have areas prone to water accumulation and areas that drain well.

Direct surface runoff can be modified by channel storage (Horton, 1937), and while watersheds might have the same drainage density, channel storage capacity might be different (Horton, 1945). The higher the channel storage ratio, the greater length of larger stream channels that can manage increased channel storage per unit of drainage area, and watersheds therefore drains out slowly (Horton, 1945). Natural storage in lakes and ponds could also initially delay flooding (Langbein, 1947). However, while buffering the yearly floods it could potentially also increase the moderate to larger floods downstream due to increased roughness from sediment loss into the lake (Ma et al., 2022) e.g. WPC and SFC with lakes closer to the mouth of the stream, and therefore would affect larger part of the runoff (Langbein, 1947). SFC is however influenced by fine glacial sediments downstream from the upper lake due to a small glacial remnant in the watershed, which can reduce flow resistance (Ma et al., 2022). The storage of stream channel reaches is often used as an index of the timing and shape of flood waves at successive points along the stream (Carter and Godfrey, 1960), since channel storage has a modulating effect in reducing flood intensity (Horton, 1937), and the larger the channel cross section, the higher the channel storage per unit length (Horton, 1945). This channel storage could result in less coarse woody debris (CWD) movement downstream, and higher retention (Naiman et al., 2002, see Klaar et al. this issue). The highest channel storage ratio was at BBS, potentially increasing the stream storage capacity compared to other streams and was the stream with the highest amount of observed CWD. SFC and WPC had the lowest channel storage ratio and CWD and would therefore potentially drain quicker. Mean bifurcation ratio is the mean of the bifurcation ratio for each stream order, where bifurcation ratio is the ratio of the number of streams of the next highest order (Horton, 1945). High bifurcation ratio indicates mountainous or highly dissected drainage basins (Horton, 1945), and the bifurcation ratio is one of the factors controlling the rate of discharge (McCullagh, 1978). This high mean bifurcation ratio indicates a shorter lag time and higher peak discharge potentially leading to floods (see Monaghan and Milner and Eagle et al. both this issue).

Area-altitude relations is important to process and patterns within the watersheds and provides a way to estimate mean snow depth or its water equivalent across the watershed (Langbein, 1947) or calculating sediment loads (Strahler, 1952). Median elevation is usually slightly below the mean elevation and probably a better indicator of various hydrologic conditions (Horton, 1932). The strong influence of topography and elevation was seen with the difference in mean and median elevations of the watershed, where the median correlated significant or stronger with other variables than the mean in relation to watershed development, and the other way around with regards to slope. Matthews and Whittaker (1987) found plant species distribution patterns demonstrate replacement of pioneer colonisers at various rates and by different species depending on local environmental factors, especially elevation, aspect and microtopology. Vegetation cover was found in areas with low slopes, and SFC and RPC which had the lowest values in relation slope and elevation and had larger areas bellow 50 m of elevation. These watersheds also supported the highest drainage density, which tend to vary with slope (Langbein, 1947). This influence also provides an indication that the initial location and topology play a significant role in watershed development. Drainage density, stream frequency and fineness ratio correlated positively with vegetation cover, showing vegetation cover influenced the drainage development by stabilisation and creation of topsoil, and may strongly regulate fluvial processes and morphology (Hickin, 1984), and in turn provide positive feedback to stream succession. Vegetation cover was negatively correlated with mean watershed slope, median elevation and length of overland flow, indicating less vegetation due to topography, and less overland flow where vegetation was present. Elevation and aspect, parent materials and degree of slope influence air and ground temperature, moisture, nutrients and materials (Swanson et al., 1988). Main channel azimuth gives an indicative location of the watershed and the legacy of glaciation and has a positive and significant correlation with relief ratio. The azimuth is also an indication of how initial location influence development, since it correlated negatively with age and total number of streams, stream order and channel storage ratio. Most morphology of stream networks are strongly influenced by innated forms (Abrahams, 1984). Drainage density is an indication of drainage development (Langbein, 1947), and high drainage density can be considering mature and low drainage density can be indication of young watersheds (Horton, 1945). The youngest stream (SFC) supported the highest drainage density and strongly indicated that development was influenced by the initial conditions. While stream development (total number of streams) was significantly related to stream age and seem to be an ongoing process, possibly due to the inherent link to precipitation and were less influenced by other processes.

The variables analysed often have a relationship with each other or are a part of equations to determine other variables. To visualise the variation within the chronosequence are the variables reduced into a two-dimensional PCA. The location of the watersheds in PCA space indicate how the different variables influence the watersheds, and the length of the arrow indicate contribution and how much variance is explained. The location also says whether close variables are correlated, no correlation at 90° and negatively correlated when opposite of each other. The variables contribution is indicated by the direction of the arrow to each of the two principal components (Figure 10), where PC1 is more a function of spatial influence (e.g area-altitude relations) while PC2 is more related to the temporal influence (e.g. age relations). PC1 show area-altitude relations with drainage density and related variables to contribute the most and suggesting that area-altitude was the most important factor when it comes to watershed development. PC2 show temporal relation with stream age as most important and then the variables that change after deglaciation, like the likelihood of loss of waterbodies, increase of stream order, denudation and erosion of the watershed and increase of bifurcation ratio, total number of streams and channel storage ratio. The importance of azimuth is related to how the placement of the watershed influences the temporal aspect. These two principal components explain 83% of the variance and indicate that the spatial aspect is more important than the temporal within the chronosequence of Glacier Bay, and that the initial conditions influence the rate of change. Since BBS and RPC are located next to each other, similar variables (e.g. geology, soil, vegetation, microclimate) might occur and therefore also they are likely to have a more similar evolutionary history. The initial watershed area-altitude conditions after the barren landscape surface from under the ice in Glacier Bay seems to influence watershed development more than age.

With more streams analysed, correlation and significance between the variables will be higher. The correlation and PCA, show that the basic variables (e.g. slope, drainage density and vegetation) are linked to the initial conditions of the watershed after deglaciation, and that the temporal effect is essentially different depending on the initial conditions. However, erosion, denudation and strong glacial isostatic adjustments are more continuous temporal processes influencing the relief and the total number of streams and showed a closer relation to a continuous development with time (e.g. stream age). Geomorphologic characteristics will influence how the stream develop and how interactions will occur e.g. difference in how floods will occur, both in intensity and duration, and will determine the affected area. These differences could be buffered by lakes and storage capacity, vegetation, soil, geology etc., all which are controlled by the initial conditions of the watershed. The findings show the temporal effect varied depending on watershed geomorphology, where development occurred at a faster rate at lower elevations, and that the initial conditions influenced the rate of colonisation and succession (Chapin et al., 1994). The initial conditions, topology, shape and history of the watershed influenced the characteristics found, and these will again influence the physical habitat and its directional development. This may change the watershed, and again change the process and patterns influencing the watershed. Based on these findings we create a conceptual framework model PHT at a point in time (PHT,t) (Fig. 11).

4.2 Framework for a remote sensed PHT at a point in time

Landforms are dependent on structure, processes and time (Davis, 1899), and the scale of investigation determines the range of patterns and processes that can be detected (Wiens, 1989). For effective research and conservation, Fausch et al. (2002) proposed a continuous view of rivers and their environment and not just of disjunct reaches, since the central problem is relating phenomena across scales (Levin, 1992). Habitat as a template for ecological strategies was proposed by Southwood (1977), and made up of a continuum of habitat heterogeneity, employing space and time as two basic dimensions. This habitat templet accommodates hypotheses development and the ordering of knowledge along spatiotemporal axes (Minshall, 1988). Minkowski (1908) formulated a "Raum-Zeitpunkt", a point in space and time, creating a four dimensional system (x,y,z,t), that we use here to create a watershed physical habitat template at a point in time (PHT,t). Watersheds

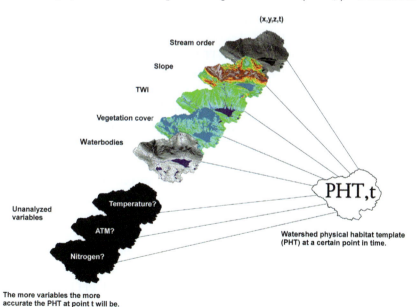

Fig. 11 Conceptual framework model for a remote sensed physical habitat template at a point in time, PHT,t.

give natural boundaries to the stream systems, and the heterogeneity within create a PHT,t as a platform for studies of riverscapes, since living beings in nature are neither distributed uniformly or at random (Legendre and Fortin, 1989). Rather than simplifying the complex environments present, the idea here is to combine collected or analysed data available into a PHT,t (Fig. 11). With the development of better remote sensing equipment, analyses of the ecological development will be improved. Cost and processing time will be the two major factors that control the choice of the investigator. Handheld versus satellite will influence the cost as well as the time needed but will also come at a different level of accuracy due to the difference in resolution. Highest resolution possible of spatially extensive multiple stream habitat components should be used to adequately represent stream ecosystem heterogeneity (McManamay and DeRolph, 2019).

While Thoms (2006) divided river ecosystems into three templates: physical, chemical and biological, they are all part of and interact in the watershed PHT,t and/or across its border. All levels within the river watershed potentially provide mechanisms to maintain biotic diversity (Medeiros et al., 2008), and the PHT,t can therefore help to understand community differences both in and among streams. Having a common PHT,t framework as a platform, datasets from different biomes across the world can be compared, and temporal aspects and climatic differences and changes analysed. This integrated approach can describe all the variables measured within a defined scale at a point in time (PHT,t), this model is set to the size of a watershed, which could change temporally. There are almost an infinite number of variables influencing the PHT,t and the more layers of variables at point t, the more robust PHT,t can be created. The PHT,t provided herein was based on measurements analysed from IFSAR remote sensing data, but there is virtually no limit on adding remote sensing layers, e.g. temperature, nitrogen, atmospheric pressure (ATM), humidity or gravity within the watershed or species distribution, biomass and so on. A three-dimensional PHT at time t create a strong basis for understanding the watershed, and can be used in all three directions, and not just a surface description as done here. Creating a point in time exist in a perfect space-temporal continuum and since not everything can be measured/analysed at ones, we must at least for now accept t as the timeframe of sampling (the period used to sample the different layers of the PHT). Using a four-dimensional PHT,t as a background for ecological interactions makes it easier to understand population colonisation, migration and interactions within the PHT, and the concept will be the same for any scale and resolution.

4.3 Directional development of the PHT,t

Crucial to the understanding of ecosystems on several temporal and spatial scales is to understand the form, behaviour, and historical context of landscapes (Swanson et al., 1988). To be able to explain changes over shorter and longer times a fourth dimension is required (Maddock, 1999). Long-term studies, which are needed to understand ecological dynamics, have to date not been satisfactorily addressed (Rull, 2014). The characteristics derived by GIS indicate differences in development within watersheds in the Glacier Bay consequence, and how the initial location of the watershed influences the direction of development in space and time. This describe the general definition of chaos as being very sensitive to initial conditions (Favis-Mortlock, 2013; Hastings et al., 1993; Phillips, 2006, 2003, 1999, 1992). The space in a four-dimensional system from point to point represents the change between them in a space-time continuum. Watershed, as a basis for a PHT,t, provides knowledge of patterns and processes within the watershed and its river system. Due to the relative short age (<250 y) difference between watersheds and minimum anthropogenic influence, Glacier Bay is a perfect natural laboratory to understand processes and patterns behind watershed and stream development (see Milner this issue). Knowing what changes and the rate of change is highly useful in understanding the processes responsible for these transitions, changes from one point to the next. The more layers that can be collated into a PHT at a point in time, the more knowledge of the interactions and processes within a watershed or the spatial scale investigated can be obtained. There will always be a close to infinity range of variables to analyse, but basic geomorphology is a solid base and likely the controlling factors in watershed development. To be able to accurately see the change from one PHT,t to another accurate georeferencing is required, otherwise watershed displacement and changes might be missed or misinterpreted. Watershed and landscape displacement continuously take place over space and time, and can occur gradually or abruptly (Gilbert, 1877), and sudden displacements are often considered as random events. Clayton et al. (2024) used Earth's elastic deformation due to water loadings to elucidate dynamic storage connectivity and watershed discharge across scales in space and time. This nonlinear dynamic displacement of a 3D integral could be the change from one PHT point in time to the next, where the spatial-temporal resolution influence if nonlinear dynamic interactions and displacements are detectable. By describing the physical laws and determine the initial conditions as precise as possible (Lorenz, 1960) changes in the PHT,t with time are best predicted.

The interconnectedness of the watershed can more readily be observed and analysed with increased resolution on a larger scale, but the larger the scale of the PHT,t the higher the demand will be for increased resolution. Each point (depending on resolution) that make up the DEM to create the PHT,t can therefore be traced from point to point in time to be able to account for the temporal and spatial change. The presented model gives a simple introduction to chaos and stochastic events in the directional development of the PHT,t. The PHT,t_1 gives a snapshot of the watersheds PHT at time t_1. The three-dimensional integral between PHT,t_1 and PHT,t_2 is the actual change in a PHT temporally from point t_1 to point t_2, and provide the rate of the changes between them. The conceptual framework model (Fig. 12) describes how deterministic chaos, and stochastic events control the direction of development, where D_{TOTAL} is the direction of development from start to end of the observed/recorded data/remotely sensed PHT at time t_6.

$$\sum_{i=1}^{N_1} D_i = \text{Sum of total directional development}$$

$$C_i = \text{Chaos}.$$

$$\sum_{j=1}^{N_2} S_{i,j} = \text{Sum of stochastic events}$$

$$D_{TOTAL} = \sum_{i=1}^{N_1} D_i = \sum_{i=1}^{N_1} \left(C_i + \sum_{j=1}^{N_2} S_{i,j} \right) = \text{Total direction of development}$$

The presented conceptual framework model describes continues nonlinear dynamic development, where process legacy from previous initial landscape conditions influence both direction, rate and the extent of change from one PHT,t point to the next. While an event might appear complex or random, it may be chaotic and thus deterministic (Wilcox et al., 1991). Thresholds are the points in which system behaviour change (Phillips, 2006). Schumm (1979) recognised both intrinsic and extrinsic geomorphic thresholds, and when passed the effect is seen in this conceptual framework model as stochastic events. This threshold is reached when change or failure occur after landform evolve to a condition of incipient instability (Schumm, 1979). Reaching and passing a threshold could be due to change in the rate of certain processes or physical laws or their extent change to lessen or increase the influence, and with each stochastic event the initial conditions changes. These nonlinear dynamic changes

Spatiotemporal dynamics 53

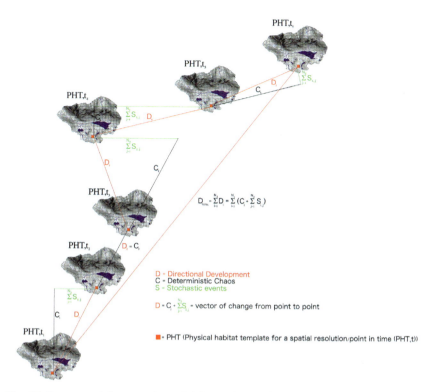

Fig. 12 Conceptual framework model for understanding directional development of the physical habitat template (PHT,t) over time.

from one point to the next can be unravelled with the help of a PHT,t, especially with the help of longer timescale monitoring or measurements of abiotic and biotic patterns and processes within the watershed PHT,t.

There is not one single direction of causality between life and its landscape; both simultaneously exert influence over each other over a range of spatial and temporal scales (Reinhardt et al., 2010). Large similarities in patterns of biodiversity, pedodiversity and geomorphological diversity suggest universal regularities in biotic and abiotic ecological structure organisation (Ibánez et al., 1995), and Scheiner and Rey-Benayas (1994) found that the landscape species richness correlated positively with community species richness. Species and communities flow, interact, and evolve through a continuum of time (Rull, 2014), and can be therefore be addressed with a PHT,t. The initial stages of biotic community development in Glacier Bay streams were determined by abiotic processes more than biotic processes (Milner, 1987), due to low biological signature right after glacial retreat. Low water temperature and channel stability in glacier fed rivers can create a habitat template that selects for certain traits, thereby suggesting deterministic processes (Milner et al., 2011, 2001). In Glacier Bay stochastic mechanisms were suggested to commence if water temperature exceeds 7 °C (Milner and Robertson, 2010). The more watersheds develop, the more feedback and complexity occur, which create more stochastic noise within the deterministic chaos. There are large variation in connectivity between real systems, from near isolation to strong mixing (Polis et al., 1997), and since recovery of disturbed sites occur deterministically through succession (Turner et al., 1993), there will be difference in interactions and feedback compared to feedback from well mixing and non-disturbed sites and again produce noise due to the difference in succession. This nonlinear dynamic chaos can produce apparently irregular fluctuations through deterministic interactions, and changes in the initial system conditions can be magnified through time (Kendall, 2001; Phillips, 1994). While each process might be deterministic, the rate and interval could be different and as result may be indistinguishable from random (Scheidegger and Langbein, 1966). Ecological interactions may not only result in spatial structures, but spatial structures may also be essential in organising the interactions (Borcard et al., 2004). Deterministic chaos can therefore be expected to have incidents of stochastic events, changing the direction of the deterministic chaos. Footprints from natural disturbances may last for a long time and can shape the ecosystem structure and function long into the future (Turner, 2010). These legacies of disturbance could be various physical structures or biological remnants from the disturbance event, e.g sediment deposition, seed transfer or input of LWD into stream systems (Foster et al., 1998), which again can change the direction of development and habitat heterogeneity. While possible random events will influence the course of colonisation and succession, initial conditions have now changed and may result in a different development direction. Complex results in chaotic systems do therefore not need to be complicated to produce (Favis-Mortlock, 2013), and random stochastic events may essentially be controlled by a complexity of nonlinear deterministic chaos.

Rapid aquatic colonisation of new systems, like in SFC (Milner et al., 2011), are found to mainly be due to watershed location and topography. Pathways are, with the use of PHT,t over time more easily detectable, and the five watersheds analysed indicated that initial conditions of the watershed to be more important than the temporal effect. No two places share identical histories, climate or topography and each have randomly fluctuating environmental variables (May, 1974). Space-for-time substitution postulate that the spatial and temporal variation is equivalent for all sites (Pickett, 1989), but we found that the studied watersheds represent various stages of development but had not traced the same evolutionary history. The use of a chronosequence in successional studies can therefore be inaccurate, particularly in relation to successional pathways of vegetation (Johnson and Miyanishi, 2008; Walker et al., 2010). Due to the different initial conditions, the watersheds will not display a direct linear development with time. A more concise method for determining change and direction of development in a chronosequence might be using a PHT,t, in order to evaluate areas that have more random changes. This directional development can help understand ecological systems where disturbance is a key component (Turner, 2010), since disturbance events might just take seconds, weeks or last for years (Burton et al., 2020). Examples could be conditions favouring colonisation of seeds transported by birds or nutrients contribution by returning Pacific salmon entering new streams for the first time. These events change the conditions of the stream system, by stabilising stream bed and contribution of nutrient to the PHT,t, and again changing the directional development of the stream system, or it could be larger disturbances like storms, floods or fires. The processes legacy (evolutionary history) behind the watershed or landscape are influenced by variables on spatial and temporal scales and can therefore be addressed with a PHT,t, since the dynamics of pristine lotic ecosystems can be understood by applying a broad spatiotemporal perspective (Ward, 1989). The greater the information collected from the entire watershed, the more of the processes and patterns in the watershed and the inhabiting biota can be analysed and assessed more fully. Due to only five study streams being studied and large differences in initial starting points after glacial retreat, it was not possible to develop a model which predicts ecological patterns and processes. Continuous PHT,t measurements will create an unprecedented resource of data to better understand changes with time. The initial conditions seem to be the strongest factor in deciding for or against colonisation and succession and influence the timeframes of both processes. The more layers that are combined in a continuum, the more temporal information can be extracted, like fluxes of temperatures, stream discharge, biomass growth, or salmonid population migration within the watershed throughout the summer. While practical impossible to analyse everything at once at time t, the rate of change will be different at different scales as well as temporally from a point in time to the next, since time is relative in relation to rate of change. However, even if the variables rate of change varies spatially and temporally from one point to the next, the sum of the change can be seen. The conceptual framework model shows how remote sensing, and GIS can help assess the rate of change in addition to the amount of change. With new technology, a better understanding of physical processes at larger spatial scales are possible. Nutrient cycling, retention and loss over time can more accurately be estimated.

The direction of watershed development based on initial conditions, chaos and the sum of stochastic events in a four dimensional space, gives the physical habitat present at point/resolution in time, 4D (PHT,t). SFC has been used throughout as an example, and in Fig. 12, a small change can be seen in the size of the watershed from PHT,t_4 to PHT,t_5, and just an illustrative way to show a small change in the size and loss of waterbodies within the watershed. There are many variables influencing the watersheds in Glacier Bay, and just one example would be the marine environment influencing the particle loading which act as cloud condensation nuclei and would influence the droplets forming and subsequently climate and rainfall in the area, and just one example of the interconnectedness within the earth's biome. Should we discuss every small aspect of influence from gravity to solar radiation we could go on forever, but the idea was to make a PHT,t where some or all these processes can be captured, and in so doing so, give valuable information about the watershed, the rate of change and the interaction between watersheds. Most models and hypothesis do not have the base to build on, and therefore will not be able to connect all the dots. However, the rate of change will be different depending on scale and resolution and different regions and for different biological connections, like energy flow and stochiometric differences in biology (e.g. plants' C3 or C4 pathways). The more watersheds develop, the more feedback and complexity occurs, which can create more noise in the deterministic chaos. The characteristics derived by GIS indicate differences in the rate and extent of change within watersheds in the Glacier Bay chronosequence, and how the initial location of the watershed influences the development direction in space and time. There is therefore no equilibrium as such, but forces continuously working on the variables in a space-time continuum.

5. Summary

Glacier Bay can function as a natural laboratory and is uniquely suited to study watershed development and interconnectedness within the landscape, due to rapid ice loss and the following colonisation and succession. Geomorphological characteristics demonstrate that the initial conditions exerts control on watershed development and influences the rate of change like show by Chapin et al. (1994), Fastie (1995), and Matthews and Whittaker (1987). The processes of stream development seem to have a more continuous rate of change, likely due to the more continuous hydrologic cycle. Deterministic chaos controls the space-time continuum of these landscapes, since a past biosphere does not exist separately from a present biosphere (Rull, 2014). Temporally the PHT,t is therefore continuously influenced by the presence of "noise" in form of stochastic events as the unseen chaos (e.g abrupt changes). We found watersheds studied herein represent various stages of development but do not share the same evolutionary history. Space-for-time substitution should therefore be used with utmost care, due to nonlinear deterministic chaos. By using a common PHT,t framework as a platform setting the scale, the facilitating variables of watersheds (e.g. morphology, climate and watershed location and so on) influence and control both colonisation and successional processes of the watershed. Since a biome is a function of their history and relation of young or mature watersheds is a function of the denudation and tectonic influence through time (short to long) (Pérez-Peña et al., 2009), a watershed will therefore never be in a state of perfect equilibrium. The watershed location within the landscape and its history in a space-time continuum like the continuous influence (e.g strong or weak) of the last Neoglacial ice age in Glacier Bay drive the evolution of the landscape. Additional research is warranted to understand the change and succession process with time, which again can help explain nutrient exchange, habitat utilisation and species and population migration patterns. The conceptual framework model should be applied with continuous remote sensing and a larger set of variables to understand the processes and patterns with time (e.g. changes and fluxes within a year). New technology and more accurate remote sensing technology contribute to a better understanding and quantification of both abiotic and biotic factors related to streams and watersheds and a more accurate PHT,t in the future. The conceptual framework PHT,t model presented here creates a more holistic understanding of the interconnectedness within a watershed and its stream system, and our findings have implications for biodiversity conservation, monitoring and management.

Acknowledgements

We thank the National Park Service (NPS) for the IFSAR data for use in GRASS 7. Thanks to Edvard T Malone, Megan Klaar, Debra Finn and Laura German for assistance in the field. Thanks to Justin Smith, Captain of the mv Capelin (NPS), for assistance to getting to the watersheds and provided fieldwork assistance.

Appendix 1. Watershed variables correlation table

	N1	N2	N3	N4	N5	N6	N7	N8	N9	N10	N11	N12	N13	N14	N15	N16	N17	N18
Age																		
Area (km2) (A)	0.300																	
Area elevation bellow 50m (km2)	0.100	-0.100																
Channel storage ratio (p)	0.700	0.300	-0.600															
Compactness	0.100	0.900*	-0.500	0.400														
Constant of channel maintenance (Ccm)	0.000	0.700	-0.600	0.300	0.900*													
Drainage density (Dd)	0.000	-0.700	0.600	-0.300	-0.900*	-1.000**												
Elongation ratio (Re)	-0.205	-0.667	0.103	-0.051	-0.667	-0.821*	0.821*											
Fineness ratio (Rf)	0.700	-0.200	0.500	0.300	-0.500	-0.700	0.700	0.462										
Form factor ratio (Ff)	-0.205	-0.667	0.103	-0.051	-0.667	-0.821*	0.821*	1.000**	0.462									
Hack main channel (km)	0.400	0.900*	-0.200	0.100	0.700	0.600	-0.600	-0.821*	-0.100	-0.821*								
Length of overland flow (Lo)	0.000	0.700	-0.600	0.300	0.900*	1.000**	-1.000**	-0.821*	-0.700	-0.821*	0.600							
Main channel azimuth	-0.900*	-0.400	0.200	-0.900*	-0.300	-0.100	0.100	0.051	-0.600	0.051	-0.300	-0.100						
Main channel sinuosity	0.100	-0.500	-0.700	0.600	-0.200	-0.100	0.100	0.462	0.100	0.462	-0.700	-0.100	-0.300					
Max watershed slope in degrees	0.100	0.400	-0.800	0.500	0.700	0.900*	-0.900*	-0.667	-0.600	-0.667	0.300	0.900*	-0.200	0.300				
Mean Bifurcation ratio (Rb)	-0.564	0.205	0.564	-0.872*	0.051	0.051	-0.051	-0.289	-0.410	-0.289	0.359	0.051	0.718	-0.872*	-0.308			
Mean Length ratio (Rl)	-0.100	0.600	-0.700	0.500	0.800	0.600	-0.600	-0.103	-0.400	-0.103	0.200	0.600	-0.300	0.200	0.500	-0.205		
Mean watershed elevation (m)	0.100	0.600	-0.300	0.100	0.700	0.900*	-0.900*	-0.975***	-0.600	-0.975***	0.700	0.900*	0.000	-0.300	0.800	0.205	0.200	
Mean watershed slope in degrees	0.000	0.700	-0.600	0.300	0.900*	1.000**	-1.000**	-0.821*	-0.700	-0.821*	0.600	1.000**	-0.100	-0.100	0.900*	0.051	0.600	0.900*

	Median watershed elevation (m)	Median watershed slope degrees	Median ruggedness number (RR)	Perimeter (km) (P)	Relief ratio (Rr)	Stream frequency (Fs)	Stream order	Total length of first order streams (km) (Ls1)	Total length of streams (km) (Ls)	Total number of streams (Ns)	Vegetation in the watershed (%)	Waterbodies (km2)	Waterbodies in the watershed (%)	Watershed length (km) (Lw)	Watershed relief (m) (H)
Median watershed elevation (m)	0.000	0.700	-0.600	0.300	0.900*	1.000**	-1.000**	-0.821*	-0.700	-0.821*	0.600	1.000**	-0.100	-0.100	0.900*	0.051	0.600	0.900*	1.000**						
Median watershed slope degrees	0.100	0.400	-0.800	0.500	0.700	0.900*	-0.900*	-0.667	-0.600	-0.667	0.300	0.900*	-0.200	0.300	1.000**	-0.308	0.500	0.800	0.900*	0.900**					
Median ruggedness number (RR)	-0.600	0.100	-0.200	-0.500	0.300	0.600	-0.600	-0.616	-0.616	0.200	0.600	0.700	-0.300	0.500	0.564	0.000	0.700	0.600	0.500						
Perimeter (km) (P)	0.300	1.000**	-0.100	0.300	0.900*	0.700	-0.700	-0.667	-0.667	0.900*	0.700	-0.400	-0.500	0.400	0.205	0.600	0.700	0.700	0.400	0.100					
Relief ratio (Rr)	-0.872*	-0.564	0.103	-0.821*	-0.410	0.154	-0.616	-0.667	0.132	-0.462	-0.154	0.975***	-0.103	-0.154	0.400	0.553	-0.359	-0.051	-0.154	-0.154	0.667	-0.564			
Stream frequency (Fs)	0.300	-0.700	0.600	-0.200	-0.900*	-0.800	0.154	0.132	-0.564	-0.400	-0.800	0.000	0.100	-0.600	-0.154	-0.600	-0.900*	-0.500	-0.800	-0.800	-0.400	-0.700			
Stream order	0.866*	0.000	-0.289	0.866*	0.000	0.000	0.800	0.410	0.410	0.700	0.000	0.000	0.577	0.100	0.289	-0.154	0.000	0.000	0.000	0.000	0.289	-0.577	0.000	0.103	
Total length of first order streams (km) (Ls1)	0.300	0.700	0.100	0.000	0.600	0.700	0.000	0.000	0.000	0.000	0.700	-0.100	-0.600	0.577	0.500	-0.889**	0.000	0.900*	0.700	0.700	0.500	0.500	0.700	-0.205	-0.300	0.000
Total length of streams (km) (Ls)	0.300	1.000**	-0.100	0.300	0.900*	0.700	-0.700	-0.667	-0.667	0.900*	0.700	-0.400	-0.500	0.400	0.205	0.600	0.600	0.700	0.700	0.700	0.400	0.100	1.000***	0.000	-0.564	-0.700	0.000
Total number of streams (Ns)	0.975***	0.205	0.051	0.667	0.051	-0.051	0.051	-0.289	-0.289	0.359	0.051	-0.821*	0.154	0.205	0.205	-0.579	-0.205	0.205	0.051	0.051	0.205	-0.462	0.100	-0.763	0.359	0.889**	0.359	0.205
Vegetation in the watershed (%)	0.400	-0.500	0.700	-0.100	-0.800	-0.900*	0.900*	0.564	0.564	-0.300	-0.900*	-0.200	0.000	0.000	-0.800	-0.154	-0.700	-0.700	-0.900*	-0.900*	-0.800	-0.700	-0.500	-0.154	0.900*	0.289	-0.400	-0.500	0.359
Waterbodies (km2)	-0.600	0.400	0.300	-0.600	0.300	0.100	0.100	-0.051	-0.051	0.300	0.100	0.500	-0.700	0.000	-0.300	0.821*	0.300	0.000	0.100	0.100	0.100	-0.300	0.300	0.400	0.308	-0.500	-0.866*	0.100	0.400	-0.718	-0.300
Waterbodies in the watershed (%)	-0.700	0.000	0.400	-0.700	-0.100	-0.300	0.300	0.359	0.359	-0.100	-0.300	0.600	-0.500	0.100	-0.600	0.718	0.100	-0.400	-0.300	-0.300	-0.600	0.100	0.000	0.462	-0.200	-0.866*	-0.300	0.000	-0.821*	0.000	0.900*
Watershed length (km) (Lw)	0.400	0.900*	0.200	0.100	0.700	0.600	-0.100	-0.821*	-0.821*	1.000**	0.600	-0.300	-0.700	0.300	0.300	0.359	0.200	0.700	0.600	0.600	0.300	0.200	0.900*	-0.462	-0.400	0.000	0.900*	0.900*	0.359	-0.300	0.300	-0.100
Watershed relief (m) (H)	0.100	0.600	-0.300	0.100	0.700	0.900*	-0.900*	-0.975***	-0.975***	0.700	0.900*	0.000	-0.300	0.800	0.205	0.200	1.000**	0.900*	0.900*	0.900*	0.800	0.700	0.600	-0.051	-0.500	0.000	0.900*	0.600	0.205	-0.700	0.000	-0.400	0.700

References

Abrahams, A.D., 1984. Channel networks: a geomorphological perspective. Water Resour. Res. 20, 161–188. https://doi.org/10.1029/WR020i002p00161.

Altamimi, Z., Gross, R., 2017. Geodesy 1039–1061. https://doi.org/10.1007/978-3-319-42928-1_36.

Blaszczynski, J.S., 1997. Landform characterization with geographic information systems. Photogrammetric Eng. Remote. Sens. 63, 183–191.

Borcard, D., Legendre, P., Avois-Jacquet, C., Tuomisto, H., 2004. Dissecting the spatial structure of ecological data at multiple scales. Ecology 85, 1826–1832. https://doi.org/10.1890/03-3111.

Burton, P.J., Jentsch, A., Walker, L.R., 2020. The ecology of disturbance interactions. BioScience 70, 854–870. https://doi.org/10.1093/biosci/biaa088.

Carbonneau, P., Fonstad, M.A., Marcus, W.A., Dugdale, S.J., 2011. Making riverscapes real. Geomorphology 137, 74–86. https://doi.org/10.1016/j.geomorph.2010.09.030.

Carter, R.W., Godfrey, R.G., 1960. Storage and flood routing (Report No. 1543B), Water Supply Paper. https://doi.org/10.3133/wsp1543B.

Chapin, F.S., Walker, L.R., Fastie, C.L., Sharman, L.C., 1994. Mechanisms of primary succession following deglaciation at Glacier Bay, Alaska. Ecol. Monogr. 64, 149. https://doi.org/10.2307/2937039.

Clayton, N., Knappe, E., White, A.M., Martens, H.R., Argus, D.F., Lau, N., et al., 2024. Elastic deformation as a tool to investigate watershed storage connectivity. Commun. Earth Environ. 5, 110. https://doi.org/10.1038/s43247-024-01264-3.

Davis, W.M., 1899. The geographical cycle. Geographical J. 14, 481. https://doi.org/10.2307/1774538.

Fassnacht, F.E., White, J.C., Wulder, M.A., Næsset, E., 2024. Remote sensing in forestry: current challenges, considerations and directions. Forestry: Int. J. For. Res. 97, 11–37. https://doi.org/10.1093/forestry/cpad024.

Fastie, C.L., 1995. Causes and ecosystem consequences of multiple pathways of primary succession at Glacier Bay, Alaska. Ecology 76, 1899. https://doi.org/10.2307/1940722.

Fausch, K.D., Torgersen, C.E., Baxter, C.V., Li, H.W., 2002. Landscapes to Riverscapes: Bridging the Gap between Research and Conservation of Stream Fishes: A Continuous View of the River Is Needed to Understand How Processes Interacting among Scales Set the Context for Stream Fishes and Their Habitat. BioScience 52, 483–498. https://doi.org/10.1641/0006-3568(2002)052[0483:LTRBTG]2.0.CO;2.

Favis-Mortlock, D., 2013. Systems and complexity in geomorphology. Treatise Geomorphology 257–270. https://doi.org/10.1016/B978-0-12-374739-6.00014-2.

Ferrari, I., Ferrarini, A., 2008. From ecosystem ecology to landscape ecology: a progression calling for a well-founded research and appropriate disillusions. Landsc. Online 6. https://doi.org/10.3097/LO.200806.

Foster, D.R., Knight, D.H., Franklin, J.F., 1998. Landscape patterns and legacies resulting from large, infrequent forest disturbances. Ecosystems 1, 497–510. https://doi.org/10.1007/s100219900046.

Frissell, C., Liss, W., Warren, C., Hurley, M., 1986. A hierarchical framework for stream habitat classification: viewing streams in a watershed context. Environ. Manag. 10, 199–214. https://doi.org/10.1007/BF01867358.

Gilbert, G.K., 1877. Report on the geology of the Henry Mountains (Report), Monograph. Washington, DC. https://doi.org/10.3133/70039916.

Glassic, H.C., McGwire, K.C., Macfarlane, W.W., Rasmussen, C., Bouwes, N., Wheaton, J.M., et al., 2024. From pixels to riverscapes: how remote sensing and geospatial tools can prioritize riverscape restoration at multiple scales. WIREs Water 11, e1716. https://doi.org/10.1002/wat2.1716.

Hack, J.T., 1957. Studies of longitudinal stream profiles in Virginia and Maryland (Report No. 294B), Professional Paper.
Hastings, A., Hom, C., Ellner, S., Turchin, P., Godfray, C., 1993. Chaos in ecology: is mother nature a strange attractor?*. Annu. Rev. Ecol. Syst. 24, 1–33. https://doi.org/10.1146/annurev.es.24.110193.000245.
Hazelton, N.W.J., Leahy, F.J., Williamson, I.P., 1990. On the design of a temporarally-referenced, 3D Geographical Information Systems: development of a four dimensional GIS. In: Proc. GIS/LIS '90, Anaheim, CA, USA, November 1990 357–372.
Hickin, E.J., 1984. Vegetation and river channel dynamics. Can. Geographer 28, 111–126.
Hohensinner, S., Jungwirth, M., Muhar, S., Schmutz, S., 2011. Spatio-temporal habitat dynamics in a changing Danube River landscape 1812—2006. River Res. Appl. 27, 939–955. https://doi.org/10.1002/rra.1407.
Horton, R.E., 1945. Erosional development of streams and their drainage basins: hydrophysical approach to quantitative morphology. Geol. Soc. Am. Bull. 56, 275. https://doi.org/10.1130/0016-7606(1945)56[275:EDOSAT]2.0.CO;2.
Horton, R.E., 1937. Natural stream channel-storage (second paper). Eos, Trans. Am. Geophys. Union. 18, 440–456. https://doi.org/10.1029/TR018i002p00440.
Horton, R.E., 1932. Drainage-basin characteristics. Eos, Trans. Am. Geophys. Union. 13, 350–361. https://doi.org/10.1029/TR013i001p00350.
Ibánez, J., Alba, S., Bermúdez Garcia-Alvarez, A., 1995. Pedodiversity: concepts and measures. Catena 24. https://doi.org/10.1016/0341-8162(95)00028-Q.
Imaizumi, F., Hattanji, T., Hayakawa, Y.S., 2010. Channel initiation by surface and subsurface flows in a steep catchment of the Akaishi Mountains, Japan. Geomorphology 115, 32–42. https://doi.org/10.1016/j.geomorph.2009.09.026.
Jasiewicz, J., Metz, M., 2011. A new GRASS GIS toolkit for Hortonian analysis of drainage networks. Computers Geosci. 37, 1162–1173. https://doi.org/10.1016/j.cageo.2011.03.003.
Jenson, S.K., 1985. Automated derivation of hydrologic basin characteristics from digital elevation model data. In: Proc. Auto-Carto 301–310.
Jenson, S.K., Domingue, J.O., 1988. Extracting topographic structure from digital elevation data for geographic information system analysis. Photogrammetric Eng. Remote. Sens. 54, 1593–1600.
Johnson, E.A., Miyanishi, K., 2008. Testing the assumptions of chronosequences in succession. Ecol. Lett. 11, 419–431. https://doi.org/10.1111/j.1461-0248.2008.01173.x.
Junk, W., Bayley, P., Sparks, R., 1989. The flood pulse concept in river-floodplain systems. Can. Spec. Public Fish. Aquat. Sci.
Kassambara, A., Mundt, F., 2020. Factoextra: extract and visualize the results of multivariate data analyses. R Package Version 1.0.7.
Kendall, B.E., 2001. Nonlinear dynamics and chaos. In: eLS. https://doi.org/10.1038/npg.els.0003314.
Khosravi, K., Melesse, A.M., Shahabi, H., Shirzadi, A., Chapi, K., Hong, H., 2019. Chapter 33—Flood susceptibility mapping at Ningdu catchment, China using bivariate and data mining techniques. In: Melesse, A.M., Abtew, W., Senay, G. (Eds.), Extreme hydrology and climate variabilityElsevier, pp. 419–434. https://doi.org/10.1016/B978-0-12-815998-9.00033-6.
Kling, G.W., 2000. Aquatic ecology: a lake's life is not its own. Nature 408, 149–150.
Langbein, W.B., 1947. Topographic characteristics of drainage basins (Report No. 968C), Water Supply Paper. https://doi.org/10.3133/wsp968C.
Legendre, P., Fortin, M.J., 1989. Spatial pattern and ecological analysis. Vegetatio 80, 107–138. https://doi.org/10.1007/BF00048036.
Levin, S.A., 1992. The problem of pattern and scale in ecology: the Robert H. MacArthur award lecture. Ecology 73, 1943–1967. https://doi.org/10.2307/1941447.

Lin, W.-T., Chou, W.-C., Lin, C.-Y., Huang, P.-H., Tsai, J.-S., 2006. Automated suitable drainage network extraction from digital elevation models in Taiwan's upstream watersheds. Hydrological Process. 20, 289–306. https://doi.org/10.1002/hyp.5911.

Lorenz, E.N., 1960. Maximum simplification of the dynamic equations. Tellus 12, 243–254. https://doi.org/10.1111/j.2153-3490.1960.tb01307.x.

Lüdecke, D., 2024. sjPlot: data visualization for statistics in social science. R Package Version 2.8.16. 2024.

Ma, H., Nittrouer, J.A., Fu, X., Parker, G., Zhang, Y., Wang, Y., et al., 2022. Amplification of downstream flood stage due to damming of fine-grained rivers. Nat. Commun. 13, 3054. https://doi.org/10.1038/s41467-022-30730-9.

Maddock, I., 1999. The importance of physical habitat assessment for evaluating river health. Freshw. Biol. 41, 373–391. https://doi.org/10.1046/j.1365-2427.1999.00437.x.

Malard, F., Tockner, K., Ward, J.V., 2000. Physico-chemical heterogeneity in a glacial riverscape. Landsc. Ecol. 15, 679–695. https://doi.org/10.1023/A:1008147419478.

Malard, F., Uehlinger, U., Zah, R., Tockner, K., 2006. Flood-pulse and riverscape dynamics in a braided glacial river. Ecology 87, 704–716. https://doi.org/10.1890/04-0889.

Mark, D.M., 1984. Part 4: Mathematical, algorithmic and data structure issues: automated detection of drainage networks from digital elevation models. Cartographica: Int. J. Geographic Inf. Geovisualization 21, 168–178. https://doi.org/10.3138/10LM-4435-6310-251R.

Marston, R.A., 2010. Geomorphology and vegetation on hillslopes: Interactions, dependencies, and feedback loops. Geomorphology 116, 206–217. https://doi.org/10.1016/j.geomorph.2009.09.028.

Matthews, J.A., Whittaker, R.J., 1987. Vegetation succession on the storbreen glacier foreland, Jotunheimen, Norway: a review. Arct. Alp. Res. 19, 385–395. https://doi.org/10.1080/00040851.1987.12002619.

Mattivi, P., Franci, F., Lambertini, A., Bitelli, G., 2019. TWI computation: a comparison of different open source GISs. Open. Geospatial Data Softw. Stand. 4, 6. https://doi.org/10.1186/s40965-019-0066-y.

May, R.M., 1974. Ecosystem patterns in randomly fluctuating environments. In: Rosen, R., Snell, F.M. (Eds.), Progress in Theoretical Biology. Academic Press, pp. 1–50.

McCullagh, P., 1978. Modern Concepts in Geomorphology. Oxford University Press.

McKean, J., Isaak, D., Wright, C., 2008. Geomorphic controls on salmon nesting patterns described by a new, narrow-beam terrestrial-aquatic lidar. Front. Ecol. Environ. https://doi.org/10.1890/070109.

McManamay, R.A., DeRolph, C.R., 2019. A stream classification system for the conterminous United States. Sci. Data 6, 190017. https://doi.org/10.1038/sdata.2019.17.

Medeiros, E., Silva, M., Ramos, R., 2008. Application of catchment- and local-scale variables for aquatic habitat characterization and assessment in the brazilian semi-arid region. Neotropical Biol. Conserv. 3(1), 13–20. ISSN 1809–9939.

Melton, M.A., 1965. The geomorphic and paleoclimatic significance of alluvial deposits in southern Arizona. J. Geol. 1–38.

Melton, M.A., 1957. An analysis of the relations among elements of climate, surface properties, and geomorphology. DTIC Document.

Mertes, L.A.K., 2002. Remote sensing of riverine landscapes. Freshw. Biol. 47, 799–816. https://doi.org/10.1046/j.1365-2427.2002.00909.x.

Milner, A., 1987. Colonization and ecological development of new streams in Glacier Bay National Park, Alaska. Freshw. Biol. 18, 53–70. https://doi.org/10.1111/j.1365-2427.1987.tb01295.x.

Milner, A.M., Brittain, J.E., Castella, E., Petts, G.E., 2001. Trends of macroinvertebrate community structure in glacier-fed rivers in relation to environmental conditions: a synthesis. Freshw. Biol. 46, 1833–1847.

Milner, A.M., Fastie, C.L., Chapin, F.S., Engstrom, D.R., Sharman, L.C., 2007. Interactions and linkages among ecosystems during landscape evolution. BioScience 57, 237–247.

Milner, A.M., Knudsen, E.E., Soiseth, C., Robertson, A.L., Schell, D., Phillips, I.T., et al., 2000. Colonization and development of stream communities across a 200-year gradient in Glacier Bay National Park, Alaska, U.S.A. Can. J. Fish. Aquat. Sci. 57, 2319–2335. https://doi.org/10.1139/f00-212.

Milner, A.M., Robertson, A.L., 2010. Colonization and succession of stream communities in Glacier Bay, Alaska; What has it contributed to general successional theory? River Res. Appl. 26, 26–35. https://doi.org/10.1002/rra.1325.

Milner, A.M., Robertson, A.L., Brown, L.E., Sønderland, S.H., McDermott, M., Veal, A.J., 2011. Evolution of a stream ecosystem in recently deglaciated terrain. Ecology 92, 1924–1935.

Minkowski, H., 1908. Die Grundgleichungen für die elektromagnetischen Vorgänge in bewegten Körpern. Nachrichten von der Gesellschaft der Wissenschaften zu Göttingen. Mathematisch-Physikalische Kl. 53–111.

Minshall, G.W., 1988. Stream ecosystem theory: a global perspective. J. North. Am. Benthological Soc. 7, 263–288. https://doi.org/10.2307/1467294.

Montgomery, D.R., Dietrich, W.E., 1992. Channel initiation and the problem of landscape scale. Science 255, 826–830. https://doi.org/10.1126/science.255.5046.826.

Montgomery, D.R., Dietrich, W.E., 1988. Where do channels begin? Nature 336, 232–234. https://doi.org/10.1038/336232a0.

Moore, I.D., Gessler, P.E., Nielsen, G.A., Peterson, G.A., 1993. Soil Attribute Prediction Using Terrain Analysis. Soil Sci. Soc. Am. J. 57, NP. https://doi.org/10.2136/sssaj1993.572NPb.

Moore, I.D., Grayson, R.B., Ladson, A.R., 1991. Digital terrain modelling: a review of hydrological, geomorphological, and biological applications. Hydrological Process. 5, 3–30. https://doi.org/10.1002/hyp.3360050103.

Mutzner, R., Tarolli, P., Sofia, G., Parlange, M.B., Rinaldo, A., 2016. Field study on drainage densities and rescaled width functions in a high-altitude alpine catchment. Hydrological Process. 30, 2138–2152. https://doi.org/10.1002/hyp.10783.

Naiman, R., Balian, E., Bartz, K., Bilby, R., Latterell, J., 2002. Dead Wood Dynamics in Stream Ecosystems. USDA Forest Service General Technical Report 181.

O'Callaghan, J.F., Mark, D.M., 1984. The extraction of drainage networks from digital elevation data. Computer Vision, Graph. Image Process. 28, 323.

Orr, H.G., Large, A.R.G., Newson, M.D., Walsh, C.L., 2008. A predictive typology for characterising hydromorphology. Geomorphology 100, 32–40. https://doi.org/10.1016/j.geomorph.2007.10.022.

Pérez-Peña, J., Azañón, J., Booth-Rea, G., Azor, A., Delgado, J., 2009. Differentiating geology and tectonics using a spatial autocorrelation technique for the hypsometric integral. J. Geophys. Res. 114. https://doi.org/10.1029/2008JF001092.

Phillips, J.D., 2006. Evolutionary geomorphology: thresholds and nonlinearity in landform response to environmental change. Hydrol. Earth Syst. Sci. 10. https://doi.org/10.5194/hessd-3-365-2006.

Phillips, J.D., 2003. Sources of nonlinearity and complexity in geomorphic systems. Prog. Phys. Geogr. 27, 1–23. https://doi.org/10.1191/0309133303pp340ra.

Phillips, J.D., 1999. Earth Surface Systems. Complexity, Order and Scale.

Phillips, J.D., 1992. Qualitative chaos in geomorphic systems, with an example from wetland response to sea level rise. J. Geol. 100, 365–374. https://doi.org/10.1086/629638.

Phillips, J.D., 2007. The perfect landscape. Geomorphology 84, 159–169. https://doi.org/10.1016/j.geomorph.2006.01.039.

Phillips, J.D., 1994. Deterministic uncertainty in landscapes. Earth Surf. Process. Landf. 19, 389–401. https://doi.org/10.1002/esp.3290190502.
Pickett, S.T.A., 1989. Long-Term Stud. Ecol. 110–135. https://doi.org/10.1007/978-1-4615-7358-6_5.
Poff, N.L., Ward, J.V., 1990. Physical habitat template of lotic systems: recovery in the context of historical pattern of spatiotemporal heterogeneity. Environ. Manag. 14, 629–645. https://doi.org/10.1007/BF02394714.
Polis, G.A., Anderson, W.B., Holt, R.D., 1997. Toward an integration of landscape and food web ecology: the dynamics of spatially subsidized food webs. Annu. Rev. Ecol. Evol. Systematics. https://doi.org/10.1146/annurev.ecolsys.28.1.289.
Poole, G., 2002. Fluvial landscape ecology: addressing uniqueness within the river discontinuum. Freshw. Biol. 47, 641–660. https://doi.org/10.1046/j.1365-2427.2002.00922.x.
Pringle, C.M., Naiman, R.J., Bretschko, G., Karr, J.R., Oswood, M.W., Webster, J.R., et al., 1988. Patch dynamics in lotic systems: the stream as a mosaic. J. North. Am. Benthological Soc. 7, 503–524. https://doi.org/10.2307/1467303.
R Core Team, 2024. R: A Language and Environment for Statistical Computing. R Foundation for Statistical Computing, URL https://www.R-project.org/. Vienna, Austria.
Reinhardt, L., Jerolmack, D., Cardinale, B.J., Vanacker, V., Wright, J., 2010. Dynamic interactions of life and its landscape: feedbacks at the interface of geomorphology and ecology. Earth Surf. Process. Landf. 35, 78–101. https://doi.org/10.1002/esp.1912.
Rocchini, D., Foody, G.M., Nagendra, H., Ricotta, C., Anand, M., He, K.S., et al., 2013. Uncertainty in ecosystem mapping by remote sensing. Computers Geosci. 50, 128–135. https://doi.org/10.1016/j.cageo.2012.05.022.
Romme, W.H., Despain, D.G., 1989. Historical perspective on the yellowstone fires of 1988. BioScience 39, 695–699. https://doi.org/10.2307/1311000.
Rull, V., 2014. Time continuum and true long-term ecology: from theory to practice. Front. Ecol. Evol. 2. https://doi.org/10.3389/fevo.2014.00075.
Sassolas-Serrayet, T., Cattin, R., Ferry, M., 2018. The shape of watersheds. Nat. Commun. 9, 3791. https://doi.org/10.1038/s41467-018-06210-4.
Scheidegger, A.E., Langbein, W.B., 1966. Probability concepts in geomorphology (Report No. 500C), Professional Paper. https://doi.org/10.3133/pp500C.
Scheiner, S.M., Rey-Benayas, J.M., 1994. Global patterns of plant diversity. Evol. Ecol. 8, 331–347. https://doi.org/10.1007/BF01238186.
Schiesari, L., Matias, M.G., Prado, P.I., Leibold, M.A., Albert, C.H., Howeth, J.G., et al., 2019. Towards an applied metaecology. Perspect. Ecol. Conserv. 17, 172–181. https://doi.org/10.1016/j.pecon.2019.11.001.
Schindler, K., Papasaika-Hanusch, H., Schütz, S., et al., 2011. Improving wide-area DEMs through data fusion -chances and limits. Photogrammetric Week '11. Wichmann Herbert, pp. 159–172.
Schroeder, D., Omran, A., Bastidas Méndez, R., 2015. Automated geoprocessing workflow for watershed delineation and classification for flash flood assessment. Int. J. Geoinform. 11. pp. 31–38.
Schumm, S.A., 1979. Geomorphic thresholds: the concept and its applications. Trans. Inst. Br. Geographers 4, 485–515. https://doi.org/10.2307/622211.
Schumm, S.A., 1956. The evolution of drainage systems and slopes in badlands at Perth Amboy, New Jersey. Geol. Soc. Am. Bull. 67, 597–646. https://doi.org/10.1130/0016-7606(1956)67[597:EODSAS]2.0.CO;2.
Schumm, S.A., 1954. The relation of drainage basin relief to sediment loss. In: Assemblée Générale de. Rome, 1954, Tome I. Int. Assoc. Sci. Hydrol. 216–219.
Schumm, S.A., Lichty, R.W., 1965. Time, space, and causality in geomorphology. Am. J. Sci. 263, 110–119. https://doi.org/10.2475/ajs.263.2.110.

Soininen, J., Bartels, P., Heino, J., Luoto, M., Hillebrand, H., 2015. Toward more integrated ecosystem research in aquatic and terrestrial environments. BioScience. https://doi.org/10.1093/biosci/biu216.

Sørensen, R., Zinko, U., Seibert, J., 2006. On the calculation of the topographic wetness index: evaluation of different methods based on field observations. Hydrol. Earth Syst. Sci. 10, 101–112. https://doi.org/10.5194/hess-10-101-2006.

Southwood, T.R.E., 1977. Habitat, the templet for ecological strategies? J. Anim. Ecol. 46, 337–365. https://doi.org/10.2307/3817.

Stanford, J., Ward, J., 1993. An ecosystem perspective of alluvial rivers: connectivity and the hyporheic corridor. J. N. Am. Benthol. Soc. 12, 48–60. https://doi.org/10.2307/1467685.

Stein, A., Gerstner, K., Kreft, H., 2014. Environmental heterogeneity as a universal driver of species richness across taxa, biomes and spatial scales. Ecol. Lett. 17. https://doi.org/10.1111/ele.12277.

Strahler, A.N., 1952. Hypsometric (area-altitude) analysis of erosional topography. Geol. Soc. Am. Bull. 63, 1117. https://doi.org/10.1130/0016-7606(1952)63[1117:HAAOET]2.0.CO;2.

Swanson, F.J., 1980. Geomorphology and ecosystems. In: Forests, Fresh Perspect. Ecosyst. Analysis: Proc. 40th Annu. Biol. Colloq. 1979, 159–170.

Swanson, F.J., Kratz, T.K., Caine, N., Woodmansee, R.G., 1988. Landform effects on ecosystem patterns and processes. BioScience 92–98.

Talluto, M., del Campo, R., Estévez, E., Altermatt, F., Datry, T., Singer, G., 2024. Towards (better) fluvial meta-ecosystem ecology: a research perspective. npj Biodivers. 3, 3. https://doi.org/10.1038/s44185-023-00036-0.

Tarboton, D.G., Ames, D.P., 2001. Advances in the mapping of flow networks from digital elevation data. In: World Water and Environmental Resources Congress. Am. Soc. Civil. Eng. USA, pp. 20–24.

Thoms, M., 2006. Variability in riverine ecosystems. River Res. Appl. 22, 115–121. https://doi.org/10.1002/rra.900.

Torgersen, C.E., Le Pichon, C., Fullerton, A.H., Dugdale, S.J., Duda, J.J., Giovannini, F., et al., 2022. Riverscape approaches in practice: perspectives and applications. Biol. Rev. 97, 481–504. https://doi.org/10.1111/brv.12810.

Toth, C., Jóźków, G., 2015. Remote sensing platforms and sensors: a survey. ISPRS J. Photogrammetry Remote. Sens. 115. https://doi.org/10.1016/j.isprsjprs.2015.10.004.

Townsend, C.R., Hildrew, A.G., 1994. Species traits in relation to a habitat templet for river systems. Freshw. Biol. 31, 265–275. https://doi.org/10.1111/j.1365-2427.1994.tb01740.x.

Turner, M., Romme, W., Gardner, R., O'Neill, R., Kratz, T., 1993. A revised concept of landscape equilibrium: disturbance and stability on scaled landscapes. Landsc. Ecol. 8, 213–227. https://doi.org/10.1007/BF00125352.

Turner, M.G., 2010. Disturbance and landscape dynamics in a changing world. Ecology 91, 2833–2849. https://doi.org/10.1890/10-0097.1.

Vannote, R.L., Minshall, G.W., Cummins, K.W., Sedell, J.R., Cushing, C.E., 1980. The river continuum concept. Can. J. Fish. Aquat. Sci. 37, 130–137. https://doi.org/10.1139/f80-017.

Walker, L.R., Wardle, D.A., Bardgett, R.D., Clarkson, B.D., 2010. The use of chronosequences in studies of ecological succession and soil development. J. Ecol. 98, 725–736. https://doi.org/10.1111/j.1365-2745.2010.01664.x.

Ward, J., Malard, F., Tockner, K., 2002. Landscape ecology: a framework for integrating pattern and process in river corridors. Landsc. Ecol. 17, 35–45. https://doi.org/10.1023/A:1015277626224.

Ward, J., Stanford, J., 1983. The serial discontinuity concept of lotic ecosystems. Dyn. Lotic Ecosyst. 10.
Ward, J.V., 1989. The four-dimensional nature of lotic ecosystems. J. North. Am. Benthological Soc. 8, 2–8. https://doi.org/10.2307/1467397.
Webster, J.R., Patten, B.C., 1979. Effects of watershed perturbation on stream potassium and calcium dynamics. Ecol. Monogr. 49, 51–72. https://doi.org/10.2307/1942572.
Wiens, J.A., 2002. Riverine landscapes: taking landscape ecology into the water. Freshw. Biol. 47, 501–515. https://doi.org/10.1046/j.1365-2427.2002.00887.x.
Wiens, J.A., 1989. Spatial scaling in ecology. Funct. Ecol. 3, 385–397. https://doi.org/10.2307/2389612.
Wilcox, B.P., Seyfried, M.S., Matison, T.H., 1991. Searching for chaotic dynamics in snowmelt runoff. Water Resour. Res. 27, 1005–1010. https://doi.org/10.1029/91WR00225.
Wilson, J.P., 2011. Digital terrain modeling. Geomorphology 137, 107–121. https://doi.org/10.1016/j.geomorph.2011.03.012.
Yang, Z., Liu, X., Zhou, M., Ai, D., Wang, G., Wang, Y., et al., 2015. The effect of environmental heterogeneity on species richness depends on community position along the environmental gradient. Sci. Rep. 5, 15723. https://doi.org/10.1038/srep15723.

CHAPTER THREE

Role of riparian vegetation in colonization and succession of stream macroinvertebrates in Glacier Bay, Alaska

Elizabeth Flory[a,*], Ian Gloyne-Phillips[b], Amanda J. Veal[c], and Alexander M. Milner[c,d]

[a]Aquatic Science Inc., Juneau, AK, United States
[b]Marine Group, NIRAS UK, Liverpool, United Kingdom
[c]School of Geography, Earth and Environmental Sciences, University of Birmingham, Edgbaston, Birmingham, United Kingdom
[d]Institute of Arctic Biology, University of Alaska Fairbanks, Fairbanks, AK, United States
*Corresponding author. e-mail address: lizflory@msn.com

Contents

1. Introduction	68
2. Methods	70
2.1 Study site	70
2.2 ST versus main channel catkin packs	71
2.3 Caddisfly body size	72
2.4 Primary productivity	73
2.5 Juvenile fish	75
3. Results	75
3.1 Tributary fauna on catkins	75
3.2 Caddisfly growth experiments	84
3.3 Primary productivity	85
3.4 Juvenile fish	87
4. Discussion	88
4.1 Riparian vegetation	88
4.2 GPP	92
Appendix	94
References	96

Abstract

The influence of riparian vegetation on macroinvertebrate community structure was investigated in a recently formed stream in southeast Alaska. Several aspects of the riparian zone were examined: (1) Colonization of willow catkins, (2) growth of caddisfly larvae fed willow catkins, and (3) gross primary productivity (GPP) seasonal fluctuations compared to macroinvertebrate density. Willow catkins and leaves increased the taxonomic richness of the Wolf Point Creek (WPC) watershed by 67%. Two groups of taxa were identified—one associated with benthic algae dominated by the chironomid *Pagastia partica*, and the other associated with terrestrial plant matter dominated by Trichoptera and the chironomids *Brillia, Corynoneura,* and *Tanytarsus*. Larvae of the caddisfly *Pseudostenophylax* grew up to 3.8 mm when fed catkins in two streams in Wachusett Inlet that had scant riparian vegetation. Significant loss of body mass occurred in the same streams when catkins were absent. Caddisflies seemed less reliant on catkins at WPC than at Wachusett, perhaps due to a more developed riparian zone at WPC offering a wider range of food. GPP showed a distinct seasonal pattern, peaking in mid-summer and was strongly correlated to macroinvertebrate density and algal biomass. Photosynthesis to respiration (P/R) ratios < 1 in early summer suggest some heterotrophy while catkins are available on stream margins while P/R > 1 later in summer suggests a shift to autotrophy when algal biomass peaks. Willow catkins may be an important seasonal input before algal growth peaks, however, retention of terrestrial inputs at WPC is limited.

1. Introduction

Terrestrial plant succession has been studied extensively in Glacier Bay National Park, southeast Alaska where rapid recession of glaciers has exposed new land and provided a range of surfaces of varying age for plants to colonize and plant communities to develop. Riparian vegetation is subject to different driving influences than vegetated areas away from streams as more light is available on stream banks allowing plants to persist that might otherwise be shaded out in the forest understory. Floods can scour stream banks and set plant succession back to an earlier stage or prevent later successional stages from becoming established. In turn, riparian vegetation exerts its own influence on adjacent stream biota, particularly macroinvertebrate communities, providing food in the form of plant material that falls into the stream (Cummins et al., 1989; Graca, 2001, Klaar et al. this issue) and providing shade that may lower water temperature variations (Sweeney & Newbold, 2014), but also reducing available light for benthic algae (Sweeney, 1992; Ghermandi et al., 2009).

In the early 1990s Wolf Point Creek (WPC) watershed (Fig. 1) supported a rapidly developing riparian zone consisting of 3–4 m willows and alder insufficient to shade the main channel due to its width. In 1993 a small (<1 m wide) side tributary (ST) flowed into WPC main channel from the proximal Minnesota Ridge approximately 500 m upstream from tidewater (Fig. 1). ST flowed through a dense canopy of alder and

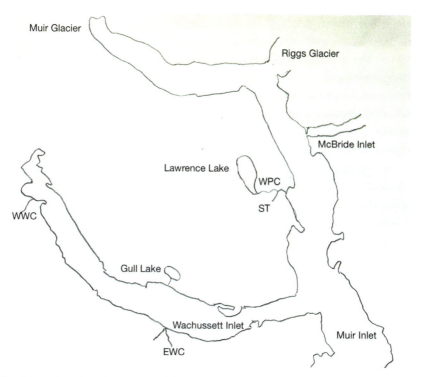

Fig. 1 Location of study streams WPC, ST, EWC and WWC in Glacier Bay, Alaska. Map lines delineate study areas and do not necessarily depict accepted national boundaries.

experienced more shading and higher retention of leaf matter from year to year than the WPC main channel. In 1993, willow catkins were observed to have fallen into the margins of WPC. Fifteen catkins from WPC near the main benthic macroinvertebrate sampling site supported 20 caddisfly larvae of varying size, not previously collected in the benthos (Flory & Milner, 1999). Willow catkins may act as an alternative energy input to the stream in summer as they are shed in June and July, unlike leaves in autumn (Flory & Milner, 1999). Further catkin macroinvertebrate assemblages were noted in ST. The potentially stronger influence of the riparian zone was examined in this study.

A comparison is made here between willow catkin packs secured in ST with identical packs secured in the main channel. Natural accumulations of leaves in ST were also collected for macroinvertebrate presence. Caddisfly larvae were previously found associated with willow catkins (Flory & Milner, 1999). Here, the importance of willow catkins to caddisfly larvae was further investigated by examining growth of larvae fed with willow catkins in WPC and two streams in nearby Wachusett Inlet.

Most streams with well-developed riparian zones receive large amounts of terrestrial carbon inputs emitting a large fraction as CO_2 to the atmosphere, similar to the net ocean CO_2 exchange (Rocher-Ros et al., 2020). WPC was characterized in the 1990s by having a limited riparian zone and extensive benthic algal growth (Adamson, 1996). Natural leaf accumulation was low in the autumn in WPC main channel, being limited to small clumps on stream margins or behind salmon carcasses when present (Flory & Milner, 1999). Estimates of total benthic gross primary productivity (GPP) are still relatively uncommon (Puts et al., 2022), but benthic resources could be the most important part of the aquatic food web in WPC. Seasonal changes in GPP and net productivity were examined in WPC in relation to changes in algal biomass and macro-invertebrate densities to examine the role of benthic algae in primary productivity. Juvenile fish rearing in the stream examined as a further variable of overall productivity.

The following hypotheses were examined:

H_1 ST will support a macroinvertebrate community more closely associated with riparian vegetation than the community in WPC owing to ST's small size under a closed canopy of alder and willow and higher retention of leaf litter than WPC.

H_2 Future development of the riparian zone of WPC and increased retention of leaf and catkin litter will result in the macroinvertebrate community in WPC containing more taxa associated with the riparian zone.

H_3 Benthic algae may be an important part of the food web for the macroinvertebrate community in WPC due to low retention of allochthonous inputs.

H_4 Willow catkins may affect the growth of caddisfly larvae.

H_5 Leaves found decaying in ST may support taxa not previously found in the watershed.

2. Methods

2.1 Study site

WPC was formed by glacial ice retreat in Muir Inlet Glacier Bay with the mouth of WPC (58.996738 N, −136.162832 W) uncovered in the mid-1940s (see Milner this issue) and aged 42 y at the time of the study and presently aged 72 y. The stream is now 2 km long fed by Lawrence Lake, also about 2 km in length. Macroinvertebrate community development in the stream has been studied since 1977. Dolly Varden (*Salvelinus malma*) colonized the stream in 1987, followed by pink (*Oncorhynchus gorbuscha*) and coho (*Onchorynchus kisutch*) salmon in 1989. Significant increases in stream temperature and decreases in turbidity were associated with decrease in glacial ice cover.

By 1993 less than 10% of the catchment was glacierized and alder (*Alnus* spp.) and willow (*Salix* spp.) were dominant with riparian plants reaching 3 m in height and pink salmon spawners numbering >10,000. In 2004, the glacial ice had virtually disappeared from the watershed and the upper terraces supported increasing numbers of cottonwood trees (*Populus trichocarpa*) along with the occasional Sitka spruce (*Picea sitchensis*). The watershed is now dominated by cottonwood with increasing abundance of Sitka spruce (Milner et al., 2018). However, the riparian zone has remained dominated by alder and willow.

2.2 ST versus main channel catkin packs

To investigate the macroinvertebrate fauna associated with willow catkins, 400 catkins were collected in July 1993 and 1994 from willow trees near WPC just before senescence. Forty "catkin packs" were constructed in 1993 and 1994 by tying 10 catkins onto 10 cm^2 of plastic-coated 5 mm wire mesh. Twenty packs were placed in the main channel within 0.5 m of the stream bank and approximately 10 m upstream of the invertebrate sampling station, which was upstream of the confluence with ST. The other 20 packs were placed in ST approximately 10 m upstream from the confluence with the main channel. Five replicate packs were retrieved from each location every 2 weeks in 1993 and every week in 1994 and the contents preserved in 70% ethanol for later analysis.

In addition, leaf packs made from dried alder leaves were placed in the main channel on August 6 1993 and 5 replicates removed each week until the early September (previously published as comparison with packs holding plastic leaves see Flory & Milner, 1999). Naturally occurring bundles of 10 leaves were collected from ST on August 5, 1994 to sample macroinvertebrate fauna on leaves as Surber sampling was not possible in ST. The relative abundance of invertebrate taxa on catkin packs and leaves in each location was compared using the ACFOR scale ranging from abundant to rare (Crisp & Southward, 1958).

Biovolume of macroinvertebrates was calculated using two 10 mL test tubes. A sample of invertebrates was firstly blotted on absorbent paper to remove excess ethanol then placed into one of the test tubes. Drops of water were slowly added to the second tube until the levels in both tubes were equal. The volume of water added to the second tube provided an estimate of invertebrate biovolume in mL.

2.3 Caddisfly body size

The extent to which catkins acted as a food source was examined using larvae of the genus *Pseudostenophylax* (Limnephilidae), commonly associated with willow catkins in ST in July 1994. *Pseudostenophylax* was the only caddisfly taxa found associated with leaf packs and the only caddis found in comparison of leaves and plastic in Flory and Milner (1999). These larvae were selected for study due to their ease of removal from their pebble cases. Eighteen *Pseudostenophylax* larvae were collected from ST on July 25, 1994. Each specimen was removed from its case and body length measured to the nearest mm. After measurement, larvae were returned to their cases and placed in one of six traps. As far as possible, three larvae of identical length were placed within each trap to allow individual growth to be examined. Individual larvae were not marked to avoid any markings potentially affecting growth rates or survival. Traps consisted of a triangular wedge-shaped wooden frame (9″ long, 6″ wide and 6″ in height) enclosed in 500 μm mesh. The pointed end of the trap was placed upstream to deflect flow and prevent mesh from becoming clogged with debris. Three caddisfly larvae were placed in each trap. Three traps were supplied with willow catkins while the remaining three were supplied with substrate covered in algae. After 14 days immersion in the stream, *Pseudostenophylax* larvae were removed and their lengths remeasured. Six traps were used in each stream to insure against the loss of some traps to potential changing flow regimes.

This method was repeated by collecting a further 36 *Pseudostenophylax* larvae on July from ST and placing them in traps in two streams in Wachusett Inlet. The stream referred to as East Wachusett Creek (EWC) enters the ocean at 58.92203 N, −136.27734 W, 8 km from Rowlee Point, where Wachusett Inlet meets Muir Inlet and drains from 1000 m elevation on Idaho Ridge (Fig. 1). The stream had no visible turbidity and daytime water temperature ranged from 7.7 °C at the beginning of the study to 9.8 °C at the end of the study. Flows ranged from 0.7 to 1 m/s and stream width averaged 5 m. Substrate consisted of large gravel with a thin layer of yellow algae. An examination of rocks with a hand lens revealed the dominance of *Diamesa davisii* chironomids.

The stream referred to as West Wachusett Creek (WWC) entered the ocean at 58.989834 N, −136.428345 W near the west end of Wachusett inlet,19.5 km from Rowlee Point, and just beyond a small (125 m by 300 m) peninsula that juts out into the terminal mudflats within 2 km (now 6 km) of the retreating Carroll Glacier (Fig. 1). *D. davisii* chironomids were found in the creek at densities of 80–100 individuals per rock. *Baetis* mayflies, Simulidae, a single megalopteran specimen (14 mm long) and two caddisfly larvae with cases constructed from small pieces of wood were also found in a pool. Daytime water temperature was 10.3 °C at the start of the study and 15.3 °C at the end while flows ranged from 0.5 to 0.7 m/s and stream width averaged 4 m.

Riparian vegetation was lacking at these stream sites in 1994, although alder was present at higher elevations. Between WPC and Wachusett Inlet are potential barriers to aerial migration in the form of the Burroughs ice remnant, the Bruce hills and Minnesota Ridge rising to nearly 1000 m. Three traps in each Wachusett stream were supplied with willow catkins from WPC while the other three traps contained only rocks with algae from the native Wachusett streams. Stream temperature at WPC ranged from 13.7 °C to 17.4 °C during the experiment.

2.4 Primary productivity

Primary productivity of WPC was estimated from 1991 to 1994 by measuring dissolved oxygen (DO) levels in paired sealed chambers placed in the stream to examine the influence of algal primary producers versus invertebrate consumers in light and dark conditions (Bott, 2007). Invertebrate densities and algal densities were also sampled and compared to oxygen production and consumption. Each chamber consisted of a 10 L volume plastic tray filled with stream water and substrate and sealed with clear plexiglass to minimize gas exchange between air and water. Eight rocks of similar size and algal biomass (and presumed macroinvertebrates) were carefully lifted from the stream bed and placed in the chambers and a submersible pump inserted to simulate stream flow. One chamber was placed inside black plastic to simulate night-time darkness while the other received natural daylight. Darkened and illuminated conditions were switched every 2 h to prevent oxygen supersaturation. DO and water temperature were measured every 2 h for a 12-h period. At the end of the sampling period, rocks were removed and individually wrapped in aluminum foil. Foil was carefully removed from overlapping areas and used to measure the surface area of the rock with the assumption that only the upper half of the rock surface was usable by both algae and invertebrates. The amount of stream water left within each tray was drained off and measured.

GPP was calculated from changes in DO levels by firstly accounting for changes in water temperature using tables of percent oxygen saturation. Adjusted changes in DO in uncovered chambers (photosynthesis) were plotted on a graph where the x-axis was time in hours. DO tends to increase from sunrise to afternoon and taper off again towards sunset. The area under this curve was calculated to estimate the daily rate of photosynthesis (Bott, 2007). Firstly the area of each trapezoid corresponding x-axis intervals in the chart of DO versus time was calculated using the formula $a = (x + y)/2 \times h$ where a = area, x = length, y = width and h = height. The area of all the trapezoids under the curve was then summed to give the total area under the curve.

Changes in DO in covered (darkened) chambers were averaged to give an hourly estimate of respiration since respiration in darkness is assumed to be more or less constant. Respiration of algae in darkness was not possible to separate from respiration of invertebrate consumers therefore daily community respiration (CR) was estimated by multiplying the hourly respiration rate by 24. Hours of daylight at the latitude of WPC ranges from 18 h in June to 15 h in August. A photoperiod of 15–18 h was used when computing CR during daylight hours to match the photoperiod of photosynthesis from the plotted graph. CR during daylight hours was computed by multiplying the hourly respiration rate by the photoperiod on the graph of DO changes.

Daily GPP estimates were then calculated by adding daily photosynthesis to CR for the photoperiod. The photosynthesis to respiration (P/R) ratio was calculated by dividing GPP by total CR over 24 h (C24). Net community metabolism was calculated by subtracting the 24 h CR from GPP. All rates were calculated per m^2 streambed substrate and volume of water within the chambers. The amount of carbon assimilated or released (net gain or loss of carbon) was calculated from net daily metabolism (NDM) using the molecular equations for O_2 to CO_2. Where photosynthesis exceed respiration, the net carbon gain was calculated using a photosynthetic quotient of 0.3125 was used, where C gained in grams = $O_2 \times 0.3125$. Where respiration exceeded photosynthesis, the net carbon loss was calculated using a respiration quotient of 0.31875, where C lost in $g = O_2 \times 0.31875$. Gross and net productivity were compared to mean densities of benthic invertebrates and algal biomass.

2.5 Juvenile fish

Twelve minnow traps baited with salmon eggs, immersed in iodine for 2 h for sterilization, were placed in small pool areas within a 100 m reach of WPC and left undisturbed for 1–1.5 h during the summers of 1993 and 1994. Any fish captured were anesthetized in MS-222 to allow identification to species and measurement of body length and weight. Fish were allowed to recover in a container of aerated stream water before being released back to the stream and mortality occurred.

3. Results
3.1 Tributary fauna on catkins

Thirty-two taxa were collected from the willow catkin packs in 1993 and 1994 from ST and the main channel combined, including 16 macroinvertebrate taxa not collected previously in main channel benthos samples. Only 27 taxa were found in benthic samples collected in 1993 and 1994 so the new taxa on catkins increased the total taxa for the WPC watershed to 42 (Table 1). Two tadpoles, thought to be boreal toads, were also found on catkins in ST, bringing the total to 43 taxa.

Twenty-nine taxa were found on ST catkin packs and 24 taxa were found on main stream catkins over the two years, although some taxa were only singly found at one sampling point. In 1993, the number of taxa found on catkins peaked at 14 after immersion for 54 d in the main channel whereas 14 taxa were found on catkins in ST after only 7 d immersion, but peaked at 18 taxa after 39 d immersion. In 1994 main channel catkins peaked at 16 taxa and ST catkins peaked at 19 taxa both after 22 d.

The new macroinvertebrate taxa found on catkins included four Limnephilid caddisfly larvae *Pyschoglypha*, *Pseudostenophylax* sp, *Ecclisommya*, and *Pycnosyche*. Fifteen caddisfly larvae specimens were collected from a single catkin in ST. Chironomids unique to catkins and present in both main channel and ST including *Pseudodiamesa branickii* Nowicki (not found in benthos in 1994), *Corynoneura* sp., *Paraphaenocladius* sp., *Limnophyes* sp. and *Paramarina* sp. Two Diptera *Pericoma* sp. (Pyschodiidae) and *Dicranota* sp. (Tipulidae) were also new to the invertebrate assemblage of the area. Five macroinvertebrate taxa unique to catkins in ST included the chironomids *Boreochlus* sp. and *Rheocricotopus* sp., the Coleopterans *Micralymma* sp. and *Celina* sp. and *Dixa* sp. (Dixidae). Only one specimen of each of the Coleoptera was found and the other taxa were present in low numbers.

Table 1 Relative abundance of taxa by microhabitat at WPC 1993-94.

Taxa	Species	Benthos	Catkins WPC	Catkins ST	Fresh leaves WPC	Decayed leaves ST
Chironmidae	Diamesa davisii	R				
	Pseudodiamesa branickii	R	R	R		R
	Pagastia partica	A	R	R	O	O
	Orthocladius manitobensis/ obumbratus	R	A	R	O	
	Orthocladius mallochi	O				O
	Paratrichocladius	R	R		R	
	Eukiefferella gracei	A	O	R	F	R
	Eukiefferella rectangularis/ Tvetania sp	R			R	
	Eukiefferella coerulescens	R			O	
	Tokunaga	R	R	R	A	
	Chaetocladius vitellinus Kieffer & Thienemann	R	F	R	O	R
	Brillia	R	R	C	C	F
	Acalcarella Shilova	R				
	Tanytarsus	R	R	F	R	R
	Telmatopelopia Fittkau	R		R		
	Corynoneura		O	C	R	R
	Paraphaenocladius		O	R	R	
	Limnophyes		O	R		
	Paramarina	R		O		R
	Boreochlus			R		
	Rheocricotopus			R		
	Thienemannia				R	O
Taxa	Species	Benthos	Catkins WPC	Catkins ST	Fresh leaves WPC	Decayed leaves ST
Ephemeroptera	Baetis rhodani	R	R	R		
Plecoptera	Neaviperla forcipata	R	F	R	R	
Trichoptera	Psychoglypha	R	R	R		F
	Pychnosyche		R	R		
	Pseudostenophylax		O	R	R	O
	Ecclisomyia		R	R		R
	Onocosmoecus	R				
Tipulidae	Dicranota		R	O		
Psychodidae	Pericoma		R	F		R
Dixidae	Dixa			R	R	
Empididae	Clinocera			R		
Muscidae	Muscidae sp A	R	R		R	
Oligochaetae	Oligochaetae sp A	R	R	O	R	C
	Oligochaetae sp B				O	
Simulidae	Prosimulium	R		R		
Coleoptera	Hydaticus		R			
	Micralymma			R		
	Celina			R		R
	Carabidae	R				
Hemiptera	Neocorixa	R				
	Number of Taxa	24	24	29	19	16
	Abundant >30%, Common 20-29%					
	Frequent 10-19%, Occasional 5-9%					
	Rare 1-4%					

None of these taxa were found in catkins in the main channel or in benthic samples. A further two taxa were present at higher abundance on catkins than in the benthos. *Clinocera* sp. (Empididae) was only present as a single specimen in the benthos in July 1994, but was collected on all but one catkin sampling occasions with mean abundance between 0.6 and 1. *Telmatopelopia* sp. (Chironomidae) was also present in the benthos at low reached a mean abundance of 3.6 in ST in August 1993, but only had a mean abundance of 1 in benthic samples on three sampling occasions between 1993 and 1994.

Species unique to the main channel catkins and not present in ST included the chironomids *Paratrichocladius* (also found in benthic samples from the main channel), a single specimen of the Muscidae and the Coleopteran *Hydaticus* sp. The chironomid *Chaetocladius*, the cranefly *Dicranota* (Tipulidae), and *Clinocera* (Empididae) were more common on catkins, but very rare in benthic samples. *Dicranota* and *Clinocera* were present on most catkin sampling occasions at mean densities of <4 and <1 per pack respectively. *Chaetocladius* reached a mean of 92 per pack on main channel catkins.

The dominant taxa present were markedly different between the two catkin locations. We collected 24 taxa in benthic samples and 29 on catkins, but only 17 of these taxa were collected in both habitats. Dominant taxa on main channel catkins were the chironomids *Orthocladius manitobensis/obumbratus, Chaetocladius* sp., *Eukiefferiella gracei, Limnophes* sp., *Paraphaenocladius* sp. and *Pagastia partica*. Few of these were present in ST at any time (Fig. 2), but on the main channel catkins *O. manitobensis* reached mean densities of 120 per pack (pp) in July 1994, *Chaetocladius* sp. reached 92 pp in August 1993, *E. gracei* reached 38 pp in August 1994 and *Limnophes* sp. reached 32 pp in August 1993. Caddisfly larvae were more abundant on catkin packs in the main channel than in ST (Fig. 3).

The dominant taxa on ST catkins were the chironomids *Brillia, Corynoneura* sp. and *Tanytarsus* sp. and *Pericoma* sp. (Pyschodidae). Additional taxa present on ST catkins at low density included the chironomid *Paramarina* sp, which was present on all sampling occasions except August 1994. *Boreochlus* sp. was found only in July 1993, while *Rheocricotopus* sp. was found only in late July and late August 1993. *Dictanota* sp. was more or less restricted to ST catkins at a density of < 4 pp with only one specimen found on main channel catkins. *Pericoma* sp. in ST comprised a large proportion of the biovolume in 1993 and 1994, along with chironomids. Caddisfly larvae made up a large proportion of the biovolume in ST in 1994 and a good portion of main channel catkins in both years.

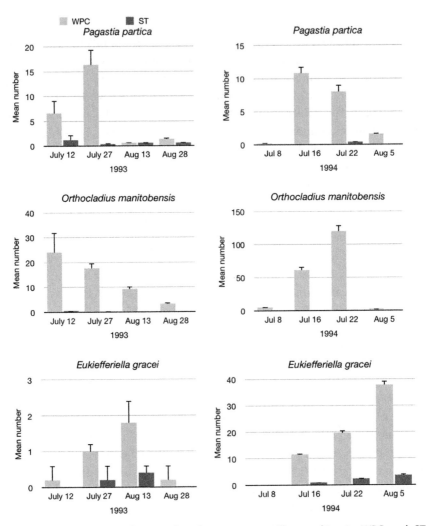

Fig. 2 Mean numbers of most abundant taxa on willow catkins in WPC and ST 1993–94.

The Venn diagram (Fig. 3) shows the amount of overlap in species presence between catkins in the two locations and the benthos. Only 12 out of 42 species collected in 1993 and 1994 were found in all three microhabitats. Catkins shared 9 taxa between locations, while 5 taxa were unique to ST catkins. Nine taxa were unique to the benthos and 15 taxa were unique to catkins.

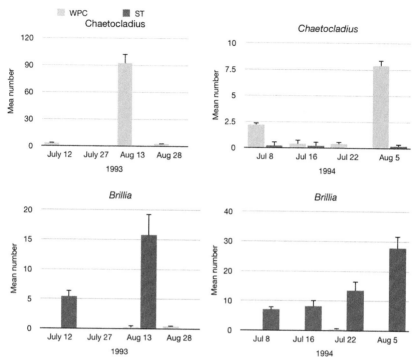

Fig. 2 (*continued*)

Natural accumulations of decaying leaves collected in 1994 in ST were found to have a different assemblage of taxa from those taxa occurring on catkins (Fig. 4). Two further uncollected taxa were the chironomid *Thienemannia* sp., and *Oligochatae* sp. B bringing the total taxa count (including tadpoles) for the watershed to 45 (1993 to 1994). The most abundant taxa on the decaying leaves collected were oligochaetes, the caddisfly *Pyschoglypha* sp. and the chironomid *Brillia* sp. Other chironomid genera present at low density were *Corynoneura* sp. *Chaetocladius, Paramarina, Pseudodiamesa braniicki, P. partica, Orthocladius mallochi, E. gracei,* and *Tanytarsus*. The Limnephilid trichopteran *Psychoglypha, Pseudostenophylax* and *Ecclisomyia* were also present at low density along with the Dipteran *Pericoma* and Coleopteran *Celina*.

Taxa present on leaf packs made from fresh dried leaves were slightly different from the decaying ones naturally present in ST. Fresh dried leaves were dominated by *Tokunaga* sp. with a mean abundance reaching 54 pp in late August 1993, *E. gracei* (21 in late August 1993), Eukiefferella*coerulescens* (9 pp), *Brillia* (8 pp), *O. manitobensis* (7 pp), *P. partica* (7 pp). Other

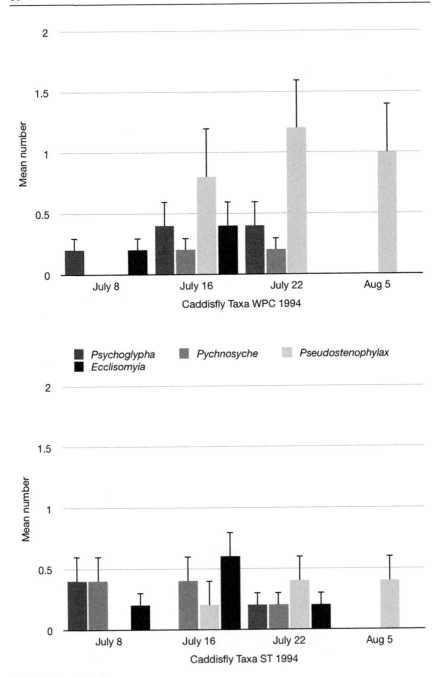

Fig. 3 (*continued*)

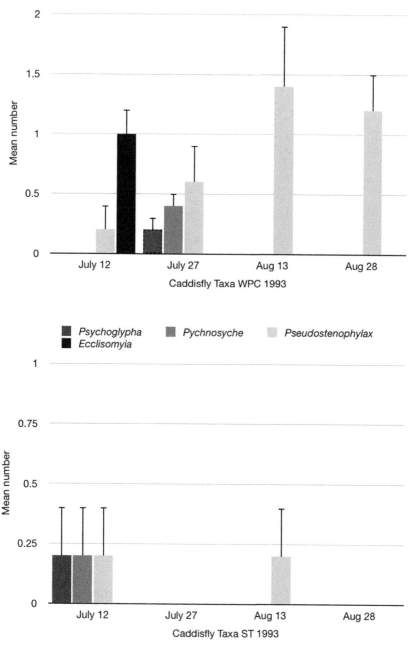

Fig. 3 Mean abundance of caddisfly larvae on willow catkins in WPC versus ST 1993. Mean abundance of caddisfly larvae on willow catkins in WPC versus ST 1994. Venn diagram depicting numbers of species present on catkins and in benthos sampled between 1993 and 1994. Numbers in overlapping circles depict number of species in common between two habitats. Only 12 species were common to all three habitats—the benthos of WPC. Nine species were unique to catkins. Forty-two species present over all three habitats.

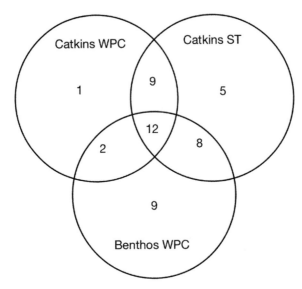

Fig. 3 (*continued*)

chironomids present at low density were *E. rectangularis/Tvetania* sp., *Tanytarsus, Paraphaenocladius, Chaetocladius, Paratrichocladius, Thienemannia* (restricted to leaves, not on catkins or in benthos). Diptera *Dixa* sp., caddisfly *Pseudostenphylax*, stonefly *Neaviperla*, simulids *Prosimulium* sp., two species of Oligochaetae and one Muscidae were also collected from fresh leaf packs. One Oligochaetae species was abundant reaching a mean density of 15 pp by September, similar to decaying leaves in ST. Natural decayed leaves compared to dried leaves appeared to have a different fauna. Table 1 compares relative abundance of taxa found in different locations. Mean abundances of all taxa are provided in Appendix 1.

The number of taxa shared between fresh leaves in WPC, decayed leaves in ST and the benthos of WPC (Fig. 4). Only 6 taxa out of 37 found were common to all three microhabitats and 9 were unique to the benthos. Thirty-seven taxa were found among catkins and leaves with 21 taxa found on both, 5 only leaves and 11 only on catkins.

Shannon index diversity and evenness are compared among the microhabitats sampled (Table 2). Decaying leaves has the greatest diversity and evenness, with a high number of species compared to total individuals sampled. Catkins had the second greatest diversity and evenness. Catkins in ST had the highest number of species compared to individuals sampled. Fresh leaves were in between catkins and decaying leaves in terms of diversity and evenness. A much greater number of individuals were

Riparian vegetation in Alaska

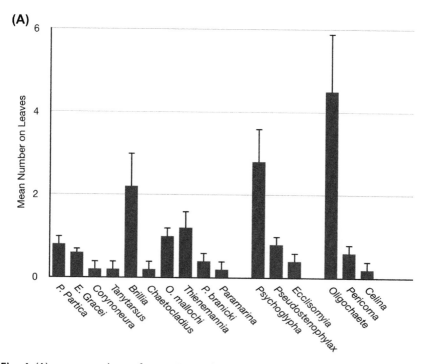

Fig. 4 (A) mean numbers of macroinvertebrates on bundles of decaying leaves collected from ST on August 5, 1994, and; (B) venn diagram showing numbers of macroinvertebrate spcies unique to each habitat and shared between habitats in 1994.

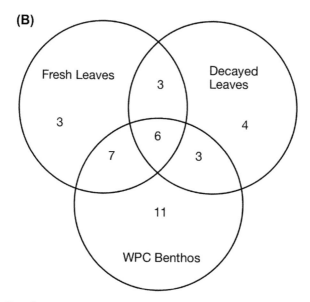

Fig. 4 (*continued*)

Table 2 Mean numbers of most abundant taxa on willow catkins in WPC and ST 1991-94.

	No. Species	Shannon	Evenness
WPC Benthos	27 (n = 7571)	1.408	0.427
WPC Catkins	24 (n = 3246)	1.834	0.577
ST Catkins	29 (n = 1639)	2.082	0.618
Fresh Leaves	19 (n = 3438)	2.135	0.725
Decaying Leaves	16 (n = 815)	2.275	0.825

Jaccard Similarity	Index
WPC Benthos vs WPC Catkins	0.52
WPC Benthos vs ST Catkins	0.52
WPC Catkins vs ST Catkins	0.72
WPC Catkins vs Decaying Leaves	0.54
ST Catkins vs Decaying Leaves	0.48
WPC Benthos vs Decaying Leaves	0.41

sampled in the benthos, but a few species (*P. partica*, and *E. gracei*) dominated giving lower diversity and evenness. The Jaccard Indices in Table 2 show that the habitats that shared the most species in common were catkins in the mainstream of WPC and catkins in ST. The least similar were decaying leaves compared to catkins or the benthos.

3.2 Caddisfly growth experiments

After two weeks, *Pseudostenphylax* supplied with catkins in all three treatment streams had maintained or slightly increased their body length, though the increases were not significant ($P < .05$). However, *Pseudostenphylax* with no access to catkins, showed a significant reduction in body mass in the two Wachusett streams ($P < .05$). Examining differences in individual *Pseudostenphylax* larvae gave insight into growth patterns with and without catkins present. Out of seven remaining caddisfly larvae fed catkins at WPC, one grew by 5 mm, two increased by 2.5 mm, two showed no change, one lost 1.3 mm and two lost 2.5 mm. Initial mean body length at WPC with catkins ranged from 15.2 to 20.3 mm declining to 17.8–19.5 mm after two weeks (Table 2). In WPC traps without catkins, five out of seven larvae remaining lost between 2.5 and 5 mm in body length and two larvae showed no change. Mean body lengths were initially 16.9–20.3 mm, but after two weeks declined to 15.2–17.8 mm. The reduction in body length was not significant.

At WWC with catkins present, two larvae gained 3.8 mm, and three gained 2.5 mm. Two larvae showed no change and two lost 0.1 mm. Where no catkins were present at WWC, two larvae were missing and the remaining seven larvae showed significant losses of between 2.5 and 10 mm ($P < .05$). At EWC with catkins present, three larvae gained 3.8 mm and one gained 2.5 mm, however, a further three larvae lost between 2.5 and 3.8 mm and two were unchanged. Changes were not significant due to the wide variation in final lengths. Where catkins were not supplied, all nine remaining larvae at EWC lost between 2.5 and 7.6 mm, a significant drop in body size ($P < .05$), despite being able to graze on algae.

No larvae grew in size without catkins present at any stream. Caddisfly larvae appeared to grow more in the two Wachussett streams than in WPC when supplied with catkins, but also lost more body mass without catkins present. Caddisfly in WPC main channel showed the smallest losses in body size without catkins.

3.3 Primary productivity

GPP, the sum of O_2 produced by photosynthesizing algae and O_2 consumed by invertebrates and other stream fauna through respiration, increased throughout the summer each year and was also higher in 1993 and 1994 than during the previous 2 years (Fig. 5). GPP remained around 100 mg O_2/m^2/day in July 1991, and July through September 1992, but reached 300 mg O_2/m^2/day in August of 1991. From late June 1993 daily GPP values exceeded 250 mg O_2/m^2. The highest GPP was recorded in late July 1993 at 519 mg O_2/m^2/day. GPP values did not reach as high in 1994, peaking at 378 mg O_2/m^2/day in August.

Net daily metabolism (NDM = respiration over 24 h subtracted from GPP) showed a net loss of carbon in 1991 and in early summer 1992–94 (Fig. 6). The biggest daily losses of NDM occurred in August 1991 (2.75 g C)

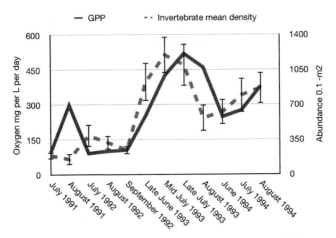

Fig. 5 Seasonal changes in GPP and macroinvertebrate density 1991–94.

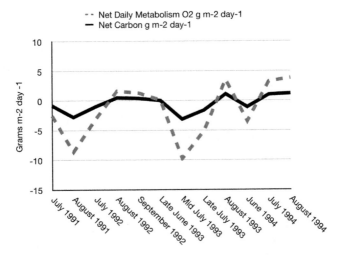

Fig. 6 Seasonal changes in net daily metabolism and net carbon.

and July 1993 (3.1 g C). Net gains occurred in August 1992 (0.5 g C), August 1993 (1.1 g C), July 1994 (1.01 g C) and August 1994 (1.2 g C). The switch from net loss to net gain appeared earlier each year from 1992 to 1994.

Increased photosynthesis and net carbon gains achieved as the summer progressed coincided with increased algal biomass, increased water temperature and decreased turbidity. Algal biomass (g) in 1994 increased steadily over the summer from around 1.8 g per 0.1 m^2 in June to 4 g in July, 7.6 g in early August and 11 g in late August. By August 1993 WPC turbidity had decreased dramatically to run clearwater some days. On August 27 1994, turbidity was measured at 14.04 NTU (Nephelometric Turbidity Units). Stream flow declined over the summer of 1994 from 4 m/s in early June to less than 0.5 m/s in early August (Adamson, 1996).

GPP was strongly correlated with macroinvertebrate density, particularly from 1992 through 1994 (Fig. 7). The r^2 value for the power of regression of GPP versus macroinvertebrate density was 0.74. A high GPP, high CR and big net loss of carbon in July 1993 coincided with very high macroinvertebrate densities (1100–1200+/−210 individuals 0.1 m^{-2}). Outliers to the curve tended to be in late summer when other factors such as algae have higher influence. Larger net losses of carbon were associated with higher invertebrate densities (linear regression r^2 = 0.7155). Net gains of carbon coincided with relatively lower densities of invertebrates at a time of year when algal biomass was greater. The ratio of daily P/R ratio only varied from 0.966 to 1.0106. A P/R ratio <1 was found in early summer each year (also in August 1991) while a P/R ratio >1 was found later in summer, but occurred earlier each year. August and September 1992, August 1993 and July and August 1994 all had P/R greater than 1. In June of 1993 P/R was just greater than 1.

3.4 Juvenile fish

Seventy-seven juvenile Dolly Varden (*S. malma*) were captured and released in WPC over 4 sampling occasions in 1993. They ranged from 7 cm fork length to 15.8 cm and from 3.5 g body weight to 45 g. Mean length was 10.6 cm +/− 1.8.

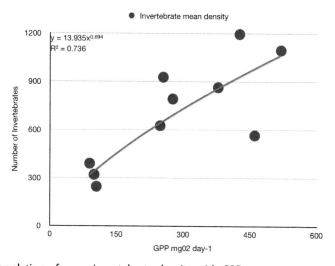

Fig. 7 Correlation of macroinvertebrate density with GPP.

Fish Condition (length/weight ratio) ranged from 0.839 to 1.313 with a mean of 1.094 +/− 0.092. Eighty-five juvenile coho salmon (*O. kisutch*) were also captured in WPC in 1993 ranging from 4.4 to 11.3 cm (mean 6.56 +/−1.43 cm). Weights ranged from 1 g for the smallest young of the year to 18.8 g for the largest smolt. Fish Condition ranged from 0.963 to 1.427 (mean 1.171 +/− 0.097). Twenty-five Dolly Varden were captured in WPC in 1994 ranging in length from 7.7 cm to 14.6 cm (mean 11.47 +/− 1.86 cm). Only five juvenile coho salmon were captured in WPC in 1994 and lengths ranged from 4.0 to 6.8 cm (mean 5.6 +/− 1.22 cm). Eighty juvenile coho salmon were captured in Gull Lake Creek, Wachusett Inlet in August 1994. Lengths ranged from 4.8 to 8.6 cm (mean 6.1 +/− 0.7 cm). Length-frequency histograms (Fig. 8) shows the numbers of fish in each size class.

4. Discussion
4.1 Riparian vegetation

The side tributary (ST) to WPC was found to support unique taxa not found in main channel benthos samples. Catkins and leaves contributed 18 taxa not previously found in benthic samples of WPC, increasing the richness of the watershed by 67%. The higher retention of catkins and leaves in ST and lower flow provided a different microhabitat that permitted different taxa to colonize. Catkins and leaves may provide a more nutritious food source, with catkins potentially constituting an important food for some caddisfly larvae. *Pseudostenophylax* caddisfly larvae appeared to rely on catkins to maintain body condition particularly in the two Wachusett streams, where algae was less abundant than at WPC. Although there was some decline in body size at WPC without catkins, the decrease was not significant. More abundant or more nutritious algae at WPC may have enabled caddisflies to maintain body mass compared to the two Wachusett streams. Small leaf particles may also have collected in traps at WPC, which had a more developed riparian zone than the Wachusett streams. WPC water temperature was also higher, reaching over 17 °C, than the two Wachusett streams perhaps affecting energy dynamics. EWC was much colder (<10 °C) compared to WWC (>15 °C). Both Wachusett streams showed large losses in body size without catkins present. Some caddisfly larvae provided with catkins in Wachusett streams showed more growth than caddisfly in WPC, so perhaps the cooler water temperature in Wachusett streams was more conducive to growth when catkins were present, but led to greater loss of

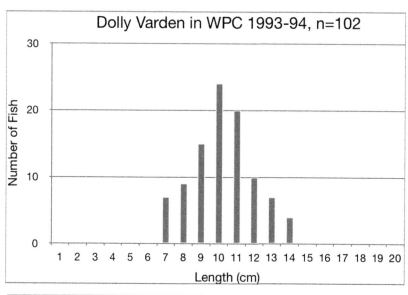

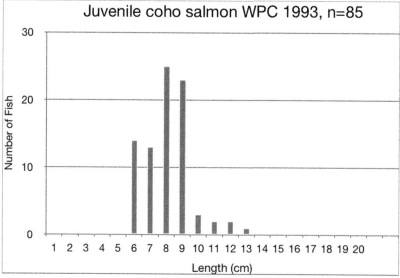

Fig. 8 Length-Frequency histograms for Dolly Varden and coho salmon in WPC 1993–94.

body mass when catkins were absent. Arsuffi and Suberkropp (1986) found that different fungal species present on leaves affected the growth of caddis larvae. The Wachusett Inlet streams may have had different fungi or algae species that colonized catkins and allowed more growth. Catkins are structurally complex (Garthwaite et al., 2021) and may have provided some protection against fluctuating flows, particularly in the small Wachusett streams that lacked a lake upstream to buffer flow variations as at WPC. Overall, catkins appeared to offer great benefit to the caddis larvae at all sites.

The caddis fly larvae *Psychoglypha*, *Pseudostenphylax* and *Ecclisomyia* were found on decaying leaf bundles. *Pseudostenophylax* was the only caddisfly taxa associated with fresh leaf packs (Flory & Milner, 1999). Also this species was the most common caddisfly on catkins and examination of guts revealed presence of plant matter in guts though they appeared to use small gravel particles for construction of cases exclusively. *Pycnopsyche* and *Psychoglypha* used leaf material for case building. A limnephelid caddisfly *Onocosmeocus* (gravel case) had been found at very low density in benthic samples at WPC, but the other four Limnephelidae were collected on catkins for the first time. The catkin food experiment suggested *Pseudostenophylax* required catkins (or a similar food) for growth. Catkins may offer higher nutritional content compared to algae. Garthwaite et al. (2021) found willow catkins were lower in tannins and higher in carbon than autumn-shed leaves and plants allocate more nutrients to reproductive parts (Wink, 2010). The chironomid *Thienemannia* was collected for the first time on decaying leaf bundles in ST adding additional taxa to the watershed. The two tadpoles found on catkins in ST may be the first documentation of amphibians utilizing plant matter deposits in a stream.

There appears to be two distinct groups of taxa present in the WPC watershed with one associated with algae, dominated by *P. partica*, and the other associated with terrestrial plant matter (e.g. catkins) dominated by caddisfly larvae and the chironomids *Brillia*, *Corynoneura* and *Tanytarsus*. The macroinvertebrate assemblage found in ST may predict future assemblages found in the main channel if retention of plant matter were to increase and the riparian zone developed enough to shade out benthic algae. But during the following 30 years however, only *Dixa* (Dixidae) appeared in the macroinvertebrate population of the benthos of the main channel. Instead of developing towards increasing proportion of fauna associated with the riparian zone, large flood events acted to reset the succession of the macroinvertebrate fauna in the stream (Milner et al., 2016).

Land surfaces in Glacier Bay recently exposed by glacial ice retreat have become colonized by shrubs and forest. The first 15–25 y of plant colonization tends to be dominated by *Dryas*, willows, cottonwood and Sitka alder (Fastie, 1995). Surfaces 35–45 y old typically have 100% cover of shrubs and young cottonwood dominated by Sitka alder. Sitka spruce is dominant on surface older than 100 years (also in closer proximity to the forest and seed source at the mouth of the bay that was untouched by ice Fastie, 1995). Succession transits from the fast-growing shade-intolerant pioneers to the slower growing shade-tolerant species. The riparian zone differs in having access to light and experiencing frequent disturbance from floods. Some riparian zones are so dynamic that succession is reset by frequent disturbance (Bendix & Hupp, 2000). If a large component of the macroinvertebrate community is associated with the riparian zone, then it is also likely kept at an earlier successional stage by floods that alter the riparian zone. Milner et al. (2018) noted that a major winter flood in 2005 reset the invertebrate community to one that resembled a community 15 years earlier. A later summer flood occurred in 2014 but the community had still not recovered from the flood 9 years earlier and the community was set back to an even earlier community.

Robertson et al. (2015) noted that the variable responses of streams in Glacier Bay to flooding events after extreme rainfall may result in differences in vegetation cover in the catchments. Older stream catchments supporting mature forests appeared more resilient to extreme rainfall events while younger streams with sparser vegetation or less complex vegetation cover were more susceptible to change. Significant changes occurred in the macroinvertebrate community in younger streams such as WPC, with total abundance and richness declining after large floods. Invertebrates more dependent on periphyton for food were the most affected by flooding (Milner et al., 2018). Communities with high proportion of Chironomidae were the most resilient to flooding. Extreme flood events are likely to increase in frequency and magnitude over the coming decades associated with climate change. Cottonwood may become more dominant in riparian zones than in areas away from streams due to light and special adaptation of cottonwood to floodplains as seedlings need continuous soil moisture and fine sediments (Eagle et al., 2021). WPC had smaller pieces of instream wood than older streams in Glacier Bay and these pieces seemed more prone to loss during floods (Klaar et al. this issue). Older streams had large instream wood that resisted substantial removal by floods. Instream wood can also be recruited by large floods that erode stream banks.

Plant litter may enter the stream from several pathways making it difficult to delineate what constitutes the riparian zone (Niaman & Decamps, 1997); overhanging branches drop leaf litter directly onto the water, lateral transport from the forest floor from rain run-off and snow melt in spring, wind-blown plant matter, and groundwater baseflow. Ultimately the extent of the riparian zone depends on the level of control the vegetation has on the stream environment. In northern southeast Alaska snow falling off trees can remove leaves and branches, which end up transported to streams during spring snowmelt and summer rains.

4.2 GPP

High levels of GPP were observed in WPC in the early 1990s in enclosed chambers. GPP magnitude was similar to studies in streams in Redwood State and National Park, California (Hill & McCormick, 2004). GPP in Peacock Creek (clearcut) reached 600 mg O_2/d while the P/R ratio was 1.13. An old growth stream in this area (Cedar Creek) was measured at 450 mg O_2/d and P/R at 0.65. Puts et al. (2022) examined total GPP in northern lakes and benthic GPP of up to 300 mg Cm^{-2}/day was found.

A higher GPP might be expected in early summer when the photoperiod extends to 18 h, but this was apparently offset or delayed by both algal biomass and invertebrate density being higher in July and August. The photoperiod dropped to 16 h by early August, but this coincided with a higher GPP in 1994. Algal biomass in 1994 accumulated over the summer as turbidity declined and water temperatures increased. The highest GPP occurred in July 1993 and coincided with high invertebrate density, particularly *P. partica*, high water temperature and the stream becoming clearwater.

GPP is comprised partly from the amount of respiration occurring during daylight hours as well as photosynthesis. A high GPP may be due to a high level of respiration or a high level photosynthesis or a combination of both. The proportion of GPP comprised of respiration tended to be highest in early summer, accounting for around 70%–75% of GPP in June and July and declining to around 65% in August. In September 1992 GPP was comprised of only 53% respiration. The remainder of the GPP was comprised of oxygen produced by photosynthesis, which increased over the summer from around 25% to 30% in June and July to 35% in August each year. The increase in photosynthesis occurred each summer despite hours of daylight decreasing from 18 h in June, to 17 in July, 16 h in August and only 13 h in September.

At WPC, GPP appeared most influenced by the amount of daylight and water temperature, both of which affect the growth of algae. P/R ratios increased over the summer as algae grew more prolific and invertebrate densities dropped. The lowest ratios occurred in June 1993 when algal biomass was low, but macroinvertebrate densities were extremely high. Willow catkins are shed into the stream at this time of the year and accumulate along stream margins, potentially offering an important food source before algal growth peaks. High macroinvertebrate densities in 1993 may have led to higher respiration rates, but small carbon gains were still observed due to high photosynthesis rates, likely from the dense mats of algae.

Stream metabolism is related to the riparian zone through inputs of plant matter to the stream and also to the amount of shading that interferes with benthic algal growth. The P/R ratio indicates whether a stream reach is a net producer or consumer of organic matter. A P/R ratio > 1 indicates a net addition of energy within the stream as there is more energy produced by algae than consumed by macroinvertebrates that is the system is autotrophic. P/R < 0.5 suggests heterotrophy while a P/R ratio > 0.5 and <1 indicates a mixture of internal and external inputs (Meyer, 1989). WPC would appear to have this mix early in the summer, but showed net gains of carbon later in summer as algae gained biomass and the stream tended towards autotrophy. In early summer there may be some use of catkins that fall into the stream leading to more heterotrophic conditions. Small fractions of catkins could have been trapped in algae or within digestive tracts of macroinvertebrates present in the sealed chambers used to measure GPP.

The use of sealed chambers only examined part of the stream ecosystem composed of algae verses macroinvertebrates. Open channel measurements would likely give very different results due to the presence of fish and other potential variables that would influence GPP. The condition factor of juvenile salmonids showed healthy fish, with high weight to length ratios suggesting they were well fed by invertebrates, also pointing to a highly productive ecosystem. Resident fish excrete potential nutrients that may be retained and utilized by algae. While retention of particulate organic material is low in WPC (Milner & Gloyne-Phillips, 2005), some dissolved organic carbon might be available and be consumed by macroinvertebrates. Nutrients like N and P from the numerous salmon carcasses observed after spawning in 1993 might have helped fuel growth of algae and macroinvertebrates, both particularly abundant that year. Previous investigations, however, did not find appreciable amounts of marine-derived N in leaves and algae (Milner et al., 2008).

There were significant changes happening in the watershed around 1993 with glacial ice in the watershed diminishing. Then stream temperatures increased and turbidity decreased to near zero and pink salmon numbers became very high. The macroinvertebrate assemblage was found to be richer than benthic samples previously showed, but subsequent flood events have kept the riparian zone at an early successional stage, perhaps slowing the advance to greater retention of leaf litter. Without establishment of larger stream bank trees there is still no source of large woody debris to trap leaf litter, but side tributaries like ST may still contribute taxa to the main channel when washed downstream, providing food for fish and enhancing the overall watershed taxa.

Appendix

Appendix 1.1: Mean abundance of taxa on willow catkins WPC versus ST 1993.

Taxa	Species	July 12 1993 WPC	July 12 1993 ST	July 27 1993 WPC	July 27 1993 ST	Aug 13 1993 WPC	Aug 13 1993 ST	Aug 28 1993 WPC	Aug 28 1993 ST
Chironomidae	Pagastia partica	6.6	1.2	16.4	0.4	0.6	0.6	1.4	
	Orthocladius manitobensis	24	0.4	17.6	0.2	9.2		3.2	
	Eukiefferiella gracei	0.2		1	0.2	1.8	0.4	0.2	
	Tokunaga	0.8		0.2		0.2	1	0.6	
	Corynoneura		0.2	0.8	9	0.4	26.8	0.2	1.6
	Tanytarsus		2		1	0.4	22.4	0.8	0.5
	Brillia		5.4			0.2	15.8	0.4	
	Paraphaenocladius							0.2	
	Chaetocladius	3.2	0.2	0.4	0.8	92.4	0.8	2.8	
	Paratrichocladius	1							
	Diamesa branickii					0.6		1.4	0.5
	Limnophes		1	0.4	1.2		0.75	31.6	
	Paramarina		1		0.4				0.5
	Boreochlus		1		0.4				
	Rheocricotopus								0.5
	Telmatopelopia						3.6		
Trichoptera	Psychoglypha		0.2	0.2					
	Pychnosyche		0.2	0.4					
	Pseudostenophylax	0.2		0.6		1.4	0.2	1.2	
	Ecclisomyia	1							
Tipulidae	Dicranota				3.8		0.6		0.7
Psychodidae	Pericoma		1		10.6	0.2	8		
Dixidae	Dixa				0.2		1.2		
Empididae	Clinocera		1		0.2		0.2		
Muscidae	Muscidae sp A								
Oligochaetae	Oligochaetae sp A				1	0.4	0.8	0.8	1
Plecoptera	Neaviperla forcipata	2		1.2		0.2			0.25
Ephemeroptera	Baetis rhodani			0.2	0.2		0.2		
Simulidae	Prosimulium		1		4				
Coleoptera	Hydaticus							0.2	
	Micralymma						0.2		
	Celina						0.2		

Appendix 1.2: Mean abundance of taxa on willow catkins WPC versus ST 1994.

Taxa	Species	July 8 1994 WPC	ST	Jul 16 1994 WPC	ST	July 22 1994 WPC	ST	Aug 5 1994 WPC	ST
Chironomidae	Pagastia partica	0.2		10.8		8	0.4	1.6	
	Orthocladius manitobensis	52		61.4		120	0.2	2.2	
	Eukiefferiella gracei			11.6	1	19.6	2.4	37.8	3.6
	Tokunaga	0.2		2		2.2	0.4		
	Corynoneura	0.4	0.6	12.2	5	12.8	21.4	0.8	28.4
	Tanytarsus		0.4	0.4	0.8	0.4	7	3.2	21.2
	Brillia		7		8.2	0.4	13.6		27.8
	Paraphaenocladius	4.6		6		19.6			0.1
	Chaetocladius	2.2	0.2	0.4	0.2	0.4		7.9	0.2
	Paratrichocladius			0.2					
	Diamesa branickii								
	Limnophes		0.2	0.2	1	0.2	0.8	0.2	0.2
	Paramarina		0.6		1	0.2	0.4		
	Boreochlus								
	Rheocricotopus								
	Telmatopelopia				0.2		0.4		
Trichoptera	Psychoglypha	0.2	0.4	0.4		0.4	0.2		
	Pychnosyche		0.4	0.2	0.4	0.2	0.2		
	Pseudostenophylax			0.8	0.2	1.2	0.4	1	0.4
	Ecclisomyia	0.2	0.2	0.4	0.6		0.2		
Tipulidae	Dicranota	0.2	0.2		1.4		0.6		0.4
Psychodidae	Pericoma		0.8		8.8		5.6		7.2
Dixidae	Dixa				0.4		1.2		0.4
Empididae	Clinocera		0.6		0.6		0.2		0.6
Muscidae	Muscidae sp A	0.2							
Oligochaetae	Oligochaetae sp A			0.6	0.8	0.4	0.4		0.4
Plecoptera	Neaviperla forcipata	0.4		4.4		2.4		1.8	
Ephemeroptera	Baetis rhodani				0.2				
Simulidae	Prosimulium								
Coleoptera	Hydaticus								
	Micralymma								
	Celina								

References

Adamson E.A.* 1996. Invertebrate Community Development in a New Stream in Glacier Bay National Park. Thesis submitted for Phd, University of Stirling, Scotland. *name change to Flory.

Arsuffi, T.L., Suberkropp, K., 1986. Growth of two stream caddisflies (Trichoptera) on leaves colonized by different fungal species. J. North. Am. Benthol. Soc., vol 5 (4), 297–305.

Bendix, J., Hupp, C.R., 2000. Hydrological and geomorphological impacts on riparian plant communities. Hydrol. Process. 14, 2977–2990.

Bott, T.L., 2007. Primary productivity and community respiration. In: Hauer, F.R., Lamberti, G. (Eds.), Chapter 28 in Methods in Stream Ecology, second ed. Elsevier.

Crisp, D.J., Southward, A.J., 1958. The distribution of intertidal organisms along the coasts of the English Channel. J. Mar. Biol. Assoc. 37, 157–203.

Cummins, K.W., Wilzbach, M.A., Gates, D.M., Perry, J.B., Talkferro, W.B., 1989. Shredders and riparian vegetation—leaf litter that falls into streams influences communities of stream invertebrates. Bioscience 39, 24–30.

Eagle, L.J.B., Milner, A.M., Klaar, M.J., Carrivick, J.L., Wilkes, M., Brown, L.E., 2021. Extreme flood disturbance effects on multiple dimensions of river invertebrate community stability. J. Anim. Ecol. 90, 2135–2146.

Fastie, C.L., 1995. Causes and ecosystem consequences of multiple pathways of primary succession at Glacier Bay, Alaska. Ecology, vol 76, 1899–1916.

Flory, E.A., Milner, A.M., 1999. Influence of riparian vegetation on invertebrate assemblages in a recently formed stream in Glacier Bay National Park, Alaska. J. North. Am. Benthol. Soc. 18 (2), 261–273.

Garthwaite, I.J., Froedin-Morgensen, A., Hartford, S.H., Claeson, S.M., Ramstack Hobbs, J.M., Leroy, C.J., 2021. Summer flower pulses: catkin litter processing in headwater streams. Fundam. Appl. Limnol. 195 (3), 243–254.

Ghermandi, A., Vandenberghe, V.L., Benedetti, L., Bauwens, W., Vanrolleghem, P.A., 2009. Model-based assessment of shading effect by riparian vegetation on river water quality. Ecol. Eng. 35 (1), 92–104.

Graca, M.A.S., 2001. The role of invertebrates in leaf litter decomposition—a review. Int. Rev. Hydrobiol. 86 (4-5), 383–393.

Hill B.H., McCormick, F.H., 2004. Nutrient uptake and community metabolism in streams draining harvested and old-growth watersheds: a preliminary assessment. In: Guldin, James M., tech. comp. 2004. Ouachita and Ozark Mountains Symposium: Ecosystem Management Research. Gen. Tech. Rep. SRS–74. U.S. Department of Agriculture, Forest Service, Southern Research Station, Asheville, NC, pp. 321.

Meyer, J.L., 1989. Can P/R ratio be used to assess the food base of stream ecosystems? A comment on Rosenfeld and Mackay (1987). Oikos 54, 119–121.

Milner, A.M., Gloyne-Phillips, I.T., 2005. The role of riparian vegetation and woody debris in the development of macroinvertebrate assemblages in streams. River Res. Appl. 21, 403–420.

Milner, A.M., Picken, J., Klaar, M.J., Robertson, A.L., Clitherow, L., Eagle, L., et al., 2018. River ecosystem resilience to extreme flood events. Ecol. Evol. 8. https://doi.org/10.1002/ece3.4300.

Milner, A.M., Robertson, A.E., Monaghan, K., Veal, A.J., Flory, E.A., 2008. Colonization and development of a stream community over 28 years; Wolf Point Creek in Glacier Bay, Alaska. Front. Ecol. Environ. 6, 413–419.

Niaman, R.J., Decamps, H., 1997. The ecology of interfaces. Annu. Rev. Ecol. Syst. 28, 621–658.

Puts, I.C., Bergström, A.K., Verheijen, Norman, H.A.S., Ask, J., 2022. An ecological and methodological assessment of benthic gross primary production in northern lakes. Ecosphere 13, e3973. https://doi.org/10.1002/ecs2.3973.

Robertson, A.L., Brown, L.E., Klaar, M.J., Milner, A.M., 2015. Stream ecosystem responses to an extreme rainfall event across multiple catchments in southeast Alaska. Freshw. Biol. 60, 2523–2534.

Rocher-Ros, G., Sponseller, R.A., Bergstrom, A.K., Myrstener, M., Giesler, R., 2020. Stream metabolism controls diel patterns and evasion of CO_2 in Arctic Streams. Glob. Change Biol. 26, 1400–1413.

Sweeney, B., 1992. Streamside forests and the physical, chemical, and trophic characteristics of Piedmont Streams in Eastern North America. Water Sci. Technol. 26 (12), 2653–2673.

Sweeney, B.W., Newbold, J.D., 2014. Streamside forest buffer width needed to protect stream water quality, habitat, and organisms: a literature review. J. Am. Water Resour. Assoc. 50 (3), 560–584.

Wink M (2010) Introduction: biochemistry, physiology and ecological functions of secondary metabolites. Annual Plant Reviews Vol 40. Biochemistry of Plant Secondary Metabolism. 2nd Ed. 1-19.

CHAPTER FOUR

Chironomidae (Diptera) community succession in streams across a 200 year gradient in Glacier Bay National Park, Alaska, USA

Alexander M. Milner[a,b,*], Katrina Magnusson[a], Amanda J. Veal[a,1], and Lee E. Brown[c]

[a]School of Geography, Earth and Environmental Sciences, University of Birmingham, Edgbaston, Birmingham, United Kingdom
[b]Institute of Arctic Biology, University of Alaska Fairbanks, Fairbanks, AK, United States
[c]School of Geography, University of Leeds, Woodhouse Lane, Leeds, United Kingdom
*Corresponding author. e-mail address: a.m.milner@bham.ac.uk

Contents

1. Introduction	100
2. Methods	101
2.1 Study site	101
2.2 Field and laboratory methods	103
3. Data analysis	105
4. Results	108
4.1 Stream habitat	108
4.2 Chironomidae community composition	108
4.3 Chironomidae community relationships with habitat variables	110
5. Discussion	113
6. Summary	115
Acknowledgments	116
Appendix	116
References	118

Abstract

In May 1997, 15 streams of contrasting stages of watershed development following glacial recession were investigated in Glacier Bay National Park, southeast Alaska, to examine chironomid community successional patterns in relation to physicochemical variables. A total of 54 taxa in five sub-families were identified; Diamesinae dominated in younger streams and Orthocladiinae in older streams. Chironominae and Tanypodinae were typically collected

[1] Current affiliation: Environment Agency, South West Region, Manley House, Kestrel Way, Exeter EX2 7LQ, UK.

only in older streams. Canonical Correspondence Analysis with species and habitat data identified four distinct stream groupings (streams < 50 years with lake influence; 50–100 years; 100–150 years and > 150 years) based upon their chironomid community composition. CBOM, stream age and presence/absence of lakes in upper reaches of streams were the most significant habitat variables influencing communities. Total abundance decreased with stream age, whereas dominance, diversity and taxon richness were similar across the stream. Clear successional patterns were evident for several taxa. *Diamesa* spp. and *Cricotopus tremulus* dominated younger streams < 100 years but were markedly less abundant in older streams. *Paratrichocladius* sp. and *Eukiefferiella rectangularis* were cosmopolitan, being found in most streams. Changes in chironomid communities over time exhibited similarities to chironomid succession found spatially along river continuums.

1. Introduction

Chironomids play an important role in the structure and function of freshwater communities due to their wide distribution and overall abundance (Cranston, 1995). Adult Chironomidae are not strong fliers but being relatively light, have high dispersal capacities (Wiederholm, 1983) permitting rapid colonization of newly formed habitat. Linked with life cycle strategies (e.g. diapause, accelerated emergence/reproduction) that allow survival in harsh conditions, chironomids are the characteristic first colonizers of disturbed stream habitats (e.g. Danks & Oliver, 1972; Ruse, 1994). Additionally, some chironomid taxa (e.g. *Diamesa*) may be better able to survive and reproduce in unstable, harsh habitats because of morphological adaptations such as modified prolegs (Milner et al., 2001) and a paucity of potential competitors and predators in these habitats (Milner, 1994; Flory & Milner, 1999). Nevertheless, Lods-Crozet et al. (2001) found high abundance of some Orthocladiinae species with *Diamesa* in unstable, harsh European glacier-fed streams, suggesting the different taxa could be spatially or temporally segregated. In contrast, some groups (e.g. Chironominae, Tanypodinae) are more typical of stable hydraulic conditions and warmer water temperature (Lingaard & Broderson, 1995; Snook & Milner, 2001).

Colonization studies of streams and rivers at the spatial scale of entirely new river channels have been limited (Fisher, 1990) and have principally involved channel relocation and reconstruction projects where upstream sources of potential drift colonizers enhance colonization rates (e.g. Gore, 1982). By using the spatial gradient of glacial recession in Glacier Bay National Park, reach scale comparisons across a temporal gradient of 200 years can be made to increase our understanding of large scale colonization patterns, rather than short-term rapid recruitment from sources within the same watershed. As these are entirely new watersheds with no remnants of any previous biological community, colonization must involve dispersal from other stream systems and development invokes primary successional processes (Gore & Milner, 1990). In a companion study, Milner et al. (2000) found significant community changes in macroinvertebrate communities in relation to physicochemical habitat variables, but Chironomidae were not identified past family level. Microcrustacea and macroinvertebrate taxa richness, and juvenile fish abundance/diversity were significantly higher in older streams. Percent contribution of Ephemeroptera also increased significantly with stream age whereas the presence of upstream lakes significantly influenced macroinvertebrate and meiofaunal abundance and percent fish cover. However, there was no significant association between stream age and total macroinvertebrate abundance due to Chironomidae being relatively abundant in the youngest streams. Although Chironomid communities play a significant role in early stream succession, we have no indication of how their community structure changes over longer successional time-scales or what are the key variables driving change.

The principal objective of this study was to investigate changes in chironomid communities across a temporal gradient of 200 years of stream development. The two principal objectives were to: (1) examine Chironomidae community composition in streams of different age and determine if distinct successional patterns exist, and; (2) understand the influence of habitat variables determining Chironomidae community composition.

2. Methods
2.1 Study site

Glacier Bay National Park and Preserve (11,030 km^2) in southeastern Alaska encompasses a fjord over 100 km long and 20 km wide with two major arms: the Northwest Arm and Muir Inlet. The maritime climate of Glacier Bay results in a mean annual temperature of 5°C (mean monthly range of −3 °C to 13 °C) and average annual precipitation of 1400 mm. Catchment characteristics can be found in Milner et al. (2000). Briefly, a Neoglacial ice sheet, which has receded from near the mouth of Glacier Bay since between 1735 and 1785 (see Fig. 1 for key recession dates and Milner this issue), has created a unique natural laboratory where scientists can study the interaction of landscape geomorphology, climate change and freshwater system development (Kling, 2000).

Fifteen streams ranging in age from 25 to 200 years since deglaciation were selected for study (Fig. 1) after Milner et al. (2000). See Sonderland &

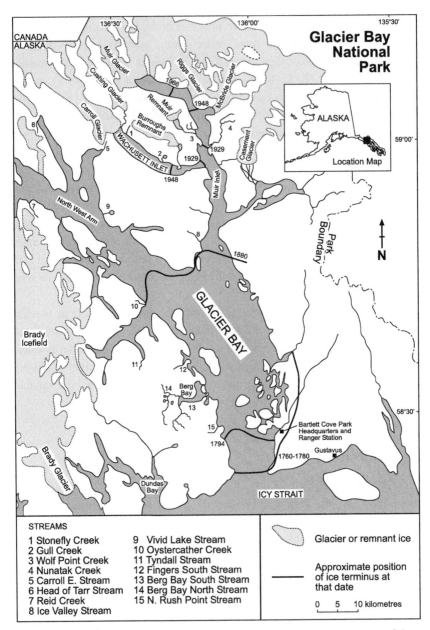

Fig. 1 Map of Glacier Bay National Park, southeast Alaska, showing locations of the 15 study streams. Map lines delineate study areas and do not necessarily depict accepted national boundaries.

Milner this issue Table 1 for extensive variables for five of these streams and Milner this issue for general glacial history. Due to insufficient information on relative deglaciation between lowland and upland portions of watersheds, stream age was defined as the time since ice recession from stream mouths in 1997 using historical and aerial photographs, journal articles, and unpublished data. All streams possessed the following characteristics: gradient <5% in the lower reaches, no barriers to salmonid migration and watershed size ranging from c. 10–100 km^2. Lakes that moderate flow variability and settle coarser sediments were present in four of the 15 streams. Proglacial lakes are frequently formed where glacial retreat forms new streams in Glacier Bay, and three of the four streams with lakes were the youngest streams in this study. The main physical features of the 15 streams are summarized in Table 1 and this is the age of the streams at the time of the study 1997. Stream order may have changed from 1997 to 2012.

2.2 Field and laboratory methods

Habitats within study reaches were identified as micro-, meso- or macro-units using a hierarchical classification system similar to that of Bryant et al. (1992). We determined *a-priori* a minimum size of > 4 m^2 for habitat classification and smaller habitats were aggregated with the habitat unit at the next highest level across the entire reach. Within each habitat classification unit, percent substrate type was visually estimated using bedrock/boulder (>256 mm), cobble (64–256 mm), gravel (2–64 mm), sand (<2 mm) or fines (silt or clay). Instream wood was estimated as the proportion of all wetted reach habitat containing any woody debris. The bottom component of the Pfankuch (1975) stability index was calculated to evaluate substrate stability in each reach (Death & Winterbourn, 1995); lower values of this index denote higher streambed stability. Water samples were collected at each site and stored as: (a) unfiltered, refrigerated; (b) unfiltered, frozen; and, (c) filtered and frozen, prior to analysis for pH, alkalinity, color and N + P nutrients. Analytical details are discussed by Milner et al. (2000). Electrical conductivity, water temperature and turbidity were recorded in the field using portable instruments.

Five random samples for invertebrates were collected from riffle areas in a 15 m section within the larger study reaches of each stream using a modified Surber sampler with 80 μm mesh. In the laboratory, chironomids were sorted using a dissecting microscope. Chironomidae were cleared in a hot solution of 10% KOH for 5–10 min then rinsed in distilled water. Chironomid head capsules were micro-dissected and mounted on slides in dimethyl-hydantoin formaldehyde resin for identification at x600–1000

Table 1 Summary of major physical watershed features of the 15 streams studied in 1997 (Stream age is in 1997).

Stream	Stream age (years)	Predominant Lake influence	Basin area (km²)	Blue line elevation (m)	Stream length (km)	Stream gradient (%)	Stream order	Orientation (° from true N)
Stonefly Creek	36	Y	~10.0	91	1.3	7.0	1	355
Gull Lake stream	43	Y	9.4	366	3.0	12.2	2	175
Wolf Point Creek	51	Y	30.8	183	5.6	3.3	2	89
Nunatak Creek	62	N	38.0	793	8.6	9.2	3	293
Carrol East stream	77	N	~10.5	518	3.4	15.2	2	226
Head of Tarr stream	88	N	8.7	579	2.8	20.7	1	332
Reid Creek	117	N	17.4	290	6.0	4.8	1	321
Head of Tyndall stream	122	N	5.7	88	2.3	3.8	2	2
Vivid Lake stream	128	N	21.6	442	5.6	7.9	2	294
Ice Valley stream	127	N	18.5	518	8.3	6.2	2	160
Oystercatcher Creek	137	N	9.6	476	5.7	8.3	2	60
North Fingers South stream	152	N	17.0	360	8.0	4.5	2	56
Berg Bay North stream	167	Y	26.8	238	9.0	2.6	3	76
Berg Bay South stream	167	N	18.6	244	7.2	3.4	3	340
Rush Point Creek	192	N	22.0	296	6.6	4.5	2	112

magnification. Particulate organic matter collected from sorted invertebrate samples was dried and weighed prior to ashing at 500°C for 40 min. CBOM was calculated as mean ash free dry mass (g) per unit area (m^{-2}).

3. Data analysis

Five biotic metrics were determined to summarize the Chironomidae communities:
1. $\log_{10} (x + 1)$ total abundance (number m^{-2});
2. number of taxa (N);
3. reciprocal of Simpson's Index (1/D) where

$$S = \frac{\sum n_i (n_i - 1)}{n(n - 1)}$$

and n_i = number of individuals in the ith species and N = the total number of individuals;
4. the Berger-Parker dominance index (d; Berger & Parker, 1970), which is a simple measure of dominance (or evenness) where
$d = N_{max}/N$
and N_{max} = the number of individuals in the most abundant taxa and N = the total number of individuals collected.
5. Percent sub families (Diamesinae, Orthocladiinae, Podominae, Chironominae, Tanypodinae)

Based upon field measurements, 19 physicochemical variables were recorded/calculated for the 15 study streams (see Milner et al., 2000 for full details of all methods). Tests for the assumption of normality revealed that half the variables were not normally distributed. Log or square root transformations were applied to the non-normal variables and the results retested for normality. Pearson's correlation coefficients were calculated to eliminate co-variables that were potential measures of the same attribute (see Table 2 for a final list of variables used in subsequent analyses).

A Detrended Correspondence Analysis (DCA) was conducted on abundance data for 54 taxa, which showed that gradients along Axis 1 were > 2.5 SD. Because gradients were long, subsequent analyses used constrained ordination methods to test for unimodal trends in compositional change at individual sites. Direct gradient analysis (Canonical Correspondence Analysis; CCA) in CANOCO 4.0 (Leps & Šmilauer, 2003) was used to determine which habitat variables accounted for a

Table 2 Mean values for physicochemical habitat variables used for analysis of relationships with chironomid communities. Asterisks denote retained variables after testing for covariance.

Variable (* denotes variables retained for further analysis)	Stonefly Creek	Gull Lake stream	Wolf Point Creek	Nunatak Creek	Carrol East stream	Head of Tarr stream	Reid Creek	Head of Tyndall stream	Vivid Lake stream	Ice Valley stream	Oystercatcher Creek	North Fingers South stream	Berg Bay North stream	Berg Bay South stream	Rush Point Creek
Turbidity (NTU)*	174	7	39	17	137	17	16	1	7	3	2	27	1	30	1
Water temperature (°C)*	2.8	3.7	3.4	5.2	5.7	2.7	3.5	3.7	6.7	8.4	2.9	3.4	4.7	6.4	6.2
Conductivity ($\mu S\ cm^{-1}$)*	115	155	101	145	181	126	148	42	133	206	129	110	170	87	170
Lake*	Y	Y	Y	N	N	N	N	N	N	N	N	N	Y	N	N
pH*	7.8	7.9	7.9	8.0	8.1	8.0	8.0	7.4	8.1	8.1	7.9	8.1	7.8	8.2	7.8
Color (Pt units)*	10	0	9	13	9	6	11	9	11	15	8	12	17	14	0
Percent Pool*	0.0	2.2	0.2	3.6	0.0	0.0	0.3	2.1	0.0	3.8	0.0	1.8	5.6	2.2	7.8
CBOM ($g\ m^{-2}$)*	0.3	0.6	0.4	0.2	0.1	0.1	0.8	0.9	0.3	0.7	0.7	0.8	0.5	1.1	1.2
Total N (ppm)*	40	2	20	60	12	20	10	0	25	173	0	197	174	99	94
Pfankuch stability index*	26	19	34	33	53	39	29	39	39	33	32	41	14	42	40
Percent bankside vegetation*	94.3	96.8	0.6	31.8	24.5	72.9	93.5	95.1	95.5	71.9	100.0	54.0	81.8	29.9	54.4
Entrenchment ratio	1.3	1.7	3.4	2.4	1.4	2.1	1.0	3.6	1.3	1.9	1.1	1.5	1.7	1.1	1.5

Percent gravel substrate	34.1	10.5	34.0	41.1	33.5	37.0	11.6	28.6	40.4	35.9	29.8	29.6	24.8	57.9	48.9
Total phosphate ($\mu g\ L^{-1}$)	53.7	3.7	16.4	127.7	6.2	49.9	8.8	248.0	33.4	24.8	3.9	1.4	2.5	47.4	11.2
Gradient	0.006	0.002	0.013	0.011	0.004	0.017	0.008	0.014	0.003	0.018	0.012	0.015	0.007	0.007	0.004
Nitrate ($\mu g\ L^{-1}$)	185.0	0.0	46.0	59.0	39.0	33.0	10.0	93.0	14.0	47.0	200.0	61.0	65.0	31.0	0.0
Alkalinity ($mg\ L^{-1}$)	49.8	87.2	51.2	42.5	93.7	69.1	58.8	55.6	51.8	63.0	54.7	18.7	80.6	39.2	60.9
Percent instream wood	12.2	34.5	22.3	14.3	7.7	6.5	5.7	1.9	0.8	5.3	4.3	16.4	22.6	6.6	14.3

significant level of the variance (forward selection; $P < 0.05$) in chironomid community data. The significance of the constrained model was tested against 499 Monte-Carlo permutations.

4. Results

4.1 Stream habitat

Habitat variables across the 15 streams are discussed fully by Milner et al. (2000). With specific reference to retained habitat variables following tests for covariance, younger streams were more turbid than older streams but the percentage of pools, CBOM, instream wood and Total N all increased significantly with stream age. Channel bed stability (Pfankuch Index) was significantly higher in reaches downstream of lakes than reaches with no upstream lakes. Stream color was higher in older streams but water temperature, turbidity, percent bankside vegetation and electrical conductivity showed no consistent temporal trends.

4.2 Chironomidae community composition

Overall, Orthocladiinae was the most abundant subfamily in the streams studied in Glacier Bay accounting for 33 of the 54 taxa but Diamesinae was dominant in six of the younger streams (Wolf Point Creek to Reid Creek along the stream age sequence and Vivid Stream; Fig. 2A). Diamesinae abundance was much lower in older streams (except Berg Bay North Stream). Chironominae (nine taxa but predominantly *Micropsectra* sp.) were relatively abundant in two of the three oldest streams and Head of Tyndall Stream. Tanypodinae were poorly represented throughout Glacier Bay. *Telmatopolopia* sp. and *Zavrelimyia* sp. were found only in Gull Lake stream (2–4 individuals m^{-2}) and *Thienemannimyia* sp. was found only in Berg Bay North Stream (>170 individuals m^{-2}). Podominae were very poorly represented being absent from all streams except Vivid Stream (*Boreochlous* sp; Fig. 2A). A list of all Chironomidae taxa found across the 15 streams is provided in Appendix Table 1.

Chironomid total abundance was approximately an order of magnitude greater in the three youngest streams and in Berg Bay North Stream (Fig. 2B); stream reaches that were influenced by lakes. Total abundance and the number of taxa showed a slight decrease from young to 100–130 year streams but then increased for streams aged 130–170 years (Fig. 3). Berger-Parker dominance and chironomid community diversity (1/Simpson's) were relatively similar across the 15 streams (Fig. 3). Peaks in dominance and

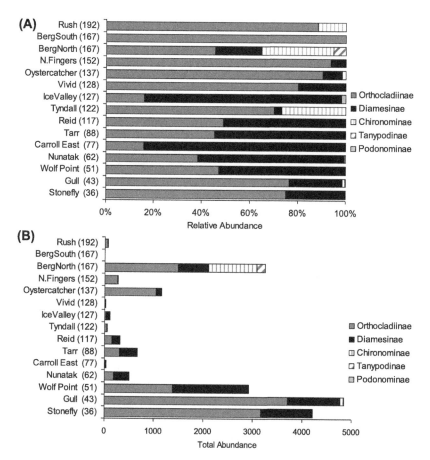

Fig. 2 (A) relative abundance, and; (B) total abundance, of chironomid sub-families in the 15 streams.

diversity for Carrol East Stream and Berg Bay South Stream corresponded to low taxon richness and abundance.

Diamesa spp. (3 species) abundance was highest in early successional streams and very few were found in older streams > 130 years (Fig. 4). In contrast, *Pagastia partica* (Diamesinae) was found in very low abundance in most streams with the exception of the older lake influenced Berg Bay North Stream. Chironominae spp were characteristic of the older streams. *Eukiefferiella* spp. (7 species) were found in all streams but were most abundant in Berg Bay South Stream. Similarly, *Orthocladius* spp. (8 species) were found in all streams but abundances were highest in Oystercatcher Creek. *Eukiefferiella claripennis* appeared to be a late successional species with highest abundance in Berg Bay

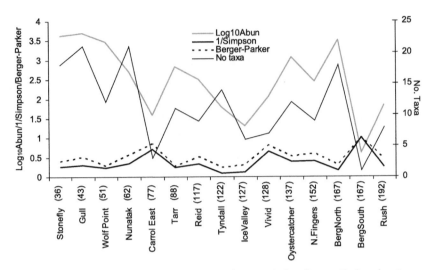

Fig. 3 Log$_{10}$ (x + 1) abundance, 1/Simpson's diversity index, Berger-Parker dominance and taxon richness for chironomid communities across the 15 streams.

South Stream, Berg Bay North Stream and North Fingers stream. *Cricotopus tremulus* was an early successional species with high relative abundance in the two youngest streams, and not found in the oldest streams except for Rush Point Creek. *Paratrichocladius* sp. was collected in most streams across the chronosequence but was most abundant in the older streams.

4.3 Chironomidae community relationships with habitat variables

Forward selection of the 11 retained habitat variables identified three significant variables (CBOM, Presence of a Lake and Stream Age) which accounted for the majority of the chironomid community variance (Fig. 5A). The first two axes of the ordination together accounted for 30.9% of the total chironomid inertia and 80% of the taxa–environment relation (Table 3). The ordination permitted the identification of four distinct stream groupings based on their chironomid community composition (Fig. 5A). The three youngest streams all aged < 50 years with lakes were grouped together and had low amounts of CBOM. Berg Bay North Stream similarly had low CBOM but had clearly different chironomid community composition than any of the four groups. The other three groups were characterized by having no lakes and were divided according to age. A second grouping of three streams (Nunatak Creek, Carrol East Stream and Head of Tarr Stream) were also relatively young (50–100 years)

Chironomidae community succession

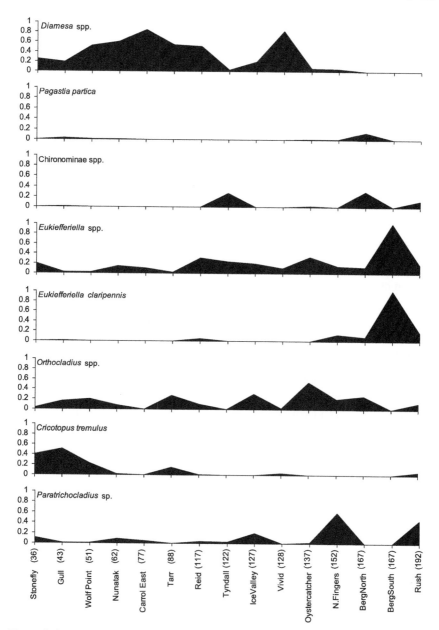

Fig. 4 Relative abundance of eight chironomid taxa across the 15 streams.

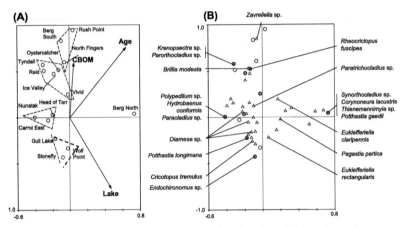

Fig. 5 Canonical Correspondence Analysis plot showing (A) location of streams in relation to forward selected habitat variables (dashed lines bound streams in different age groups; see text for discussion), and; (B) selected chironomid taxa in relation to location of streams.

Table 3 Summary statistics for canonical correspondence analysis of chironomid taxa abundances and forward selected habitat variables.

Forward selected habitat variables	Percent species total inertia along axis 1 (axes 1 & 2)	Percent species-environment relation along axis 1 (axes 1 & 2)	Significance (P)
CBOM Stream Age Lake	18.4 (30.9)	47.5 (80.0)	<0.01

with low CBOM. A third group of five streams (Oystercatcher Creek, Head of Tyndall Stream, Reid Creek, Vivid Lake Stream and Ice Valley Stream) between 100–150 years was evident with higher CBOM, and a fourth group of streams > 150 years (excluding Berg Bay North Stream) included streams with the highest levels of CBOM.

Many chironomid taxa were common to all 15 streams (i.e. located centrally within the ordination space; Fig. 5B). These included *Paratrichocladius* sp. and *E. claripennis*. However, the grouping of younger streams with lakes was strongly related to *Diamesa* spp., *Potthastia longimana* and *C. tremulus*; however, other Diamesinae species were more closely related to older streams (e.g. *Potthastia gaedii*; Berg Bay North Stream). Berg Bay North Stream contained four taxa not found in any other streams (i.e. *Synorthocladius* sp. *Corynoneura lacustris*, *Thienemannimyia* sp. and *P. gaedii*). The second stream group (50–100 years; without lakes) was strongly grouped by *Polypedilum* sp. *Hydrobaenus conformis*, *Paracladius* sp. and a *Diamesa* sp. Older streams without lakes were predominantly separated by Orthocladiinae and Chironominae taxa. Group 3, was grouped by *Brillia modesta*, *Krenopsectra* sp., *Rheocrictopus fuscipes* and *Parorthocladius* sp. Berg Bay South Stream and Rush Point Creek were grouped according to the presence of *Zavreliella* sp.

5. Discussion

Clear successional changes were evident in the chironomid communities across a 200 year chronosequence of streams in Glacier Bay, southeast Alaska. Clear differences were evident between streams of different ages, with some taxa characteristic of either young or older streams, or some taxa common to most streams. Chironomidae sub-families also changed across the chronosequence from Diamesinae dominated in younger streams, to Orthocladiinae dominated in older streams. Chironominae relative abundances were also higher in older streams. Chironomid community composition across the chronosequence was influenced predominantly by stream age, the availability of CBOM and the presence of flow buffering lakes influencing downstream reaches.

Chironomidae abundance was an order of magnitude higher (reaching ~ 4800 m^{-2}) in the stable channel of the four Glacier Bay stream channels influenced by lakes, compared to non-lake system streams where chironomid densities rarely exceeded 1000 m^{-2}. Autochthonous production is enhanced in these young, stable channels and dense mats of filamentous algae and/or moss are present on the stream bed. These mats trap fine organic particles (some of lake origin) and provide increased food supplies for Chironomidae, thereby enhancing their abundance. Overall there was an increase in CBOM with stream age as stream retention characteristics (e.g. woody debris, pool habitat) become more widespread in older streams (Milner & Phillips, 2005). This may account for the switch from grazers (*Diamesa*) in the younger stream to more collector gathers feeding on detritus in the older streams (e.g. *B. modesta*, *Zavreliella* sp.).

The four stream groups identified herein using the chironomid community dataset were similar to the TWINSPAN stream groups identified in an earlier study of the 15 streams using the entire macroinvertebrate community with Chironomidae data at family level (Milner et al., 2000). This indicates the stream groupings of post-glacial streams driven by stream age and the influence of a lake on downstream reaches are robust. In this current study, the three lake influenced systems were grouped together but Berg Bay North Stream, the oldest lake influenced system, was a distinct outlier. However, Berg Bay North Stream was located in a similar position of the ordination along axis 2 as the three other lake systems. The stream was separated along axis 1 by four chironomid taxa that were not found in other streams. This suggests that changes in taxonomic composition are evident between lake-fed systems of different ages, comparable to changes in communities of non-lake fed systems with increasing stream age.

Taxon richness, diversity and evenness of the chironomid community did not change markedly over the 15 streams and the 200 years of development. This is in marked contrast to mayfly and stonefly taxon richness and diversity, which showed a significant increase with stream age in an associated study, as a function of greater availability of coarse benthic organic matter (Milner et al., 2000). CBOM was also a major variable driving the chironomid community structure in the older streams in this study. However, chironomid communities probably displayed little change in dominance or diversity over time because many taxa within this family span the range of functional groups and possess adaptive traits for many stream habitat scenarios. Thus, different Chironomidae taxa are able to colonize a wide variety of stream types in different stages of development across landscapes from young, cold, unstable channels (e.g. *Diamesa*) to older, warmer streams with increased allochthonous inputs (e.g. Chironominae).

Chironomid community patterns across the stream chronosequence was remarkably similar to the chironomid subfamily patterns described for European streams by Lindegaard and Brodersen (1995) along a continuum of stream types from glacier fed streams to lower mountain rivers. In glacier brooks, Diamesinae are reported to have marginally higher relative abundance (56%) than Orthocladiinae (44%) with no other subfamilies present. However, in our study the two youngest systems with lakes were dominated by Orthocladiinae spp. (74%–77%) in concordance with Mackay (1992) who suggested that the first chironomid colonizers of disturbed habitats were Orthocladiinae. The predicted pattern for Diamesinae was generally similar to streams in Glacier Bay up to 117 years age excluding these two youngest streams, after which time Orthocladiinae again dominated. Lindegaard and Brodersen (1995) also suggested that relative abundance of Chironominae increased as stream conditions became more stable (i.e. along the continuum). Similarly, we found more Chironominae in older, warmer streams with better developed, more stable habitat, similar to Lindegaard and Brodersen's (1995) lower mountain classification. However, we found very few Tanypodinae in

Glacier Bay; if similar patterns to Lindegaard and Brodersen (1995) were evident, then we would have expected relative abundance to reach approximately 9% in older streams. Tanypodinae were only found in high abundance in the lake influenced Berg Bay North Stream. Tanypodinae are predators and this functional group is usually the last to colonize streams (Mackay, 1992). We suggest that older streams up to 200 years old have not yet acquired habitat conditions sufficient to support large Tanypodinae populations, except in streams where stream habitat development has been enhanced by the buffering characteristic of lakes. Nevertheless, our results suggest that spatial patterns in Chironomidae community composition might be evident temporally given sufficient scales of examination.

Our results across the stream chronosequence also bear similarity to the temporal successional pattern of Wolf Point Creek, one of the younger streams in this study but for which we now possess a > 28 year record of chironomid colonization and succession (Milner, 1994; Flory & Milner, 2000; Veal, 2004). The first colonizers of Wolf Point Creek belonged to the genus *Diamesa*; specifically a species of the *Diamesa davisi* group with densities exceeding 2900 m^{-2}. Some of the early colonizing *Diamesa* species (notably *Diamesa sommermanni*) were not collected from 1988 and *Diamesa* abundance has declined markedly as was evident across the chronosequence. *P. partica* (Diamesinae) and *Orthocladius mallochi*, first collected in Wolf Point Creek in 1986, both increased in abundance (reaching 3000 m^{-2}) to be the co-dominant chironomid species in 1994 but have since subsequently declined in abundance. *E. claripennis* and *C. tremulus*, both first collected in 1986, were co-dominant in 2001. *C. tremulus* was characteristic of younger streams in the chronosequence (streams aged < 50 years). Interestingly, *Paratrichocladius* sp. is the only chironomid that has been collected in Wolf Point Creek every year since the study began and was one of the most cosmopolitan taxa across the 15 streams, being found in young streams but also very abundant in older streams. Some chironomid taxa are cosmopolitan, whereas other taxa are early and late successional taxa. Cosmopolitan taxa are presumably able to survive harsh conditions but also tolerant of a range of habitat conditions as streams age. These taxa must also be good competitors and generalist feeders, to compete as other taxa colonize, and adapt to changes in food resources with age.

6. Summary

In summary, chironomids are one of the most widely distributed and abundant groups of benthic macroinvertebrates and this study provides new insights into how these communities are structured over long time-periods in relation to stream age, CBOM and lakes influencing stream flow. The study demonstrated clear successional patterns among the Chironomidae, from early successional communities dominated by Diamesinae taxa, to older streams dominated by Orthocladiinae but with increased abundances of Chironominae. Different successional changes over time were evident for individual taxa within the same subfamily (e.g. Diamesinae); notably the early successional genus *Diamesa* contrasted with *P. partica* which was found in older streams, suggesting a range of possible adaptations possessed by chironomids even at lower taxonomic levels (< subfamily). Community change over time in these North American streams paralleled community change along large-scale spatial continuums found for European streams. This finding suggests similarities in spatial and temporal chironomid community succession, and tentatively points towards similar Chironomidae community structure and function between continents. Further studies of similarities in spatial and temporal patterns of community structure and function (within and between continents) could contribute greatly to our understanding of large-scale patterns in stream ecology.

Acknowledgments

Eric Knudsen, Chad Soiseth, Anne Robertson, Ian Phillips, Carol Woody, Kevin Sage, John Sargent, Robert Schmeh, and Kim Scribner assisted with field and laboratory work. We are grateful to the National Park Service for supplying the m.v. Stellar, and to her crew Captain Dan Foley and first mate Rocky Wood. We thank Dr. Clive Pinder for verifying the chironomids. We are particularly grateful to the Alaska Department of Fish and Game Limnology Laboratory in Soldotna, particularly Ginny Litchfield and Jim Edmundson, for water sample analysis. AMM acknowledges the support of the UK National Environment Research Council (grant GR9/2913) for funding towards this research.

Appendix

Table 1 List of Chironomidae taxa found in 15 streams in Glacier Bay, May 1997.
Orthocladiinae

Cricotopus tremulus	
Paratrichocladius	**Tanypodinae**
Orthocladius/Cricotopus sp.4	*Telmatopelopia* sp.
Orthocladius/Cricotopus sp.6	*Zavrelimyia* sp.
Orthocladius/Cricotopus sp.7	*Thienemannimyia* sp.
Euorthocladius sp.1	*Podonominae* sp.

Orthocladius/Cricotopus sp.9	*Boreochlous* sp.
Orthocladius/Cricotopus sp.10	
Orthocladius/Cricotopus sp.15	
Orthocladius/Cricotopus sp.17	**Chironominae**
Synorthocladius sp.	*Micropsectra* sp.
Parorthocladius sp.	*Krenopsectra* sp.
Brillia modesta	*Polypedilum* sp.1
Corynoneura lobata type	*Paraclodopelma* sp.
Corynoneura lacustris type	*Sergentia* sp.
Eukiefferiella cyanea	*Zavreliella* sp.
Eukiefferiella claripennis sp.1	*Endochironomus* sp.
Eukiefferiella rectangularis sp.1	Chironominae s
Eukiefferiella courelescens	*Corynocera* sp.
Eukiefferiella brehmi	
Eukiefferiella gracei/Tvetenia sp.1	**Diamesinae**
Eukiefferiella/Tvetenia sp.2	*Diamesa* sp.1
Tvetenia sp.1	*Diamesa* sp.2
Diplocladius sp.	*Diamesa* sp.3
Camptocladius/Pseudosmittia	*Pagastia partica*
Chaetocladius acuticornis/dentiforceps grp	*Potthastia gaedii*
Hydrobaenus conformis	*Potthastia longimana*
Paracladius sp.	*Pseudokiefferiella* sp.
Parakiefferiella sp.	
Paraphaenocladius sp	
Psectrocladius sordidellus	
Rheocricotopus fuscipes	
Thienemaniella clavicornis	

References

Berger, W.H., Parker, F.L., 1970. Diversity of planktonic Foraminifera in deep sea sediments. Science 168, 1345–1347.

Bryant, M.D., Wright, B.E., Davies, B.J., 1992. Application of a Hierarchical Habitat Unit Classification System: Stream Habitat and Salmonid Distribution in Ward Creek, Southeast Alaska. U.S.D.A. Forest Serv., Pac. NW Res. Sta. Res. Note PNW-RN-508. 18 pp.

Cranston, P.S., 1995. Introduction to the chironomidae. In: Armitage, P., Cranston, P.S., Pinder, L.C.V. (Eds.), The Chironomidae—The Biology and Ecology of Non-biting Midges. Chapman and Hall, pp. 1–7.

Danks, H.V., Oliver, D.R., 1972. Seasonal emergence of some arctic Chironomidae (Diptera). Can. Entomol. 104, 661–686.

Death, R.G., Winterbourn, M.J., 1995. Diversity patterns in stream benthic invertebrates: the influence of habitat stability. Ecology 76, 1446–1460.

Fisher, S.G., 1990. Recovery processes in lotic ecosystems. Environ. Manag. 14, 725–736.

Flory, E.A., Milner, A.M., 1999. The role of competition in invertebrate community development in a recently formed stream in Glacier Bay National Park, Alaska. Aquat. Ecol. 33, 175–184.

Flory, E.A., Milner, A.M., 2000. Macroinvertebrate community succession in Wolf Point Creek, Glacier Bay National Park, Alaska USA. Freshw. Biol. 44, 465–480.

Gore, J.A., 1982. Benthic invertebrate colonization: source distance effects on community composition. Hydrobiologia 94, 183–193.

Gore, J.A., Milner, A.M., 1990. Island biogeographical theory: can it be used to predict lotic recovery rates? Environ. Manag. 14, 737–753.

Kling, 2000. A lake's life is not its own. Nature 408, 149–150.

Leps, J., Šmilauer, P., 2003. Multivariate Analysis of Ecological Data Using CANOCO. Cambridge University Press,, pp. 268.

Lods-Crozet, B., Lencioni, V., Olafson, J.S., Snook, D.L., Velle, G., Brittain, J.E., et al., 2001. Chironomid (Diptera: Chironomidae) communities in six European glacier-fed streams. Freshw. Biol. 46, 1791–1810.

Lingaard, C., Broderson, K.P., 1995. Distribution of Chironomidae (Diptera) in the river continuum. In: Cranston, P. (Ed.), Chironomids: From Genes to Ecosystems. CSIRO, Melbourne, pp. 257–271.

Mackay, R.J., 1992. Colonization by lotic macroinvertebrates: a review of processes and patterns. Can. J. Fish. Aquat. Sci. 49, 617–628.

Milner, A.M., 1994. Invertebrate colonization and succession in a new stream in Glacier Bay National Park, Alaska. Freshw. Biol. 32, 387–400.

Milner, A.M., Knudsen, E.E., Soiseth, C., Robertson, A.L., Schell, D., Phillips, I.T., et al., 2000. Colonization and development of stream communities across a 200 year gradient in Glacier Bay National Park, Alaska. Can. J. Fish. Aquat. Sci. 57, 2319–2335.

Milner, A.M., Phillips, I.T., 2005. The role of riparian vegetation and woody debris in the development of macroinvertebrate assemblages in streams. River Res. Appl. 21, 403–420.

Pfankuch, D.J., 1975. Stream Reach Inventory and Channel Stability Evaluation. United States Department of Agriculture Forest Service, Region 1, Missoula, Montana, USA.

Ruse, L.P., 1994. Chironomid microdistribution in gravel of an English chalk river. Freshw. Biol. 32, 533–551.

Snook, D.L., Milner, A.M., 2001. The influence of glacial runoff on stream macroinvertebrate communities in the Taillon catchment, French Pyrénées. Freshw. Biol. 46, 1609–1623.

Veal, A.J. 2004. Macroinvertebrate community and habitat development in Stonefly Creek; A newly formed catchment in Glacier Bay, Alaska. PhD thesis, University of Birmingham. pp. 243.

Wiederholm, T. 1983. Chironomidae of the Holarctic Region. Keys and Diagnoses. Part 1. Larvae. Entomologica Scandinavica Supplement 19: pp. 449.

CHAPTER FIVE

Salmon lice (*Lepeophtheirus salmonis*) as a food source for juvenile salmonids in Glacier Bay, southeast Alaska

Svein Harald Sønderland[a,*] and Alexander M. Milner[a,b]
[a]School of Geography, Earth and Environmental Sciences, University of Birmingham, Edgbaston, Birmingham, United Kingdom
[b]Institute of Arctic Biology, University of Alaska Fairbanks, Fairbanks, AK, United States
*Corresponding author. e-mail address: snarrald@gmail.com

Contents

1. Introduction	122
2. Methods	124
2.1 Study area	124
3. Sample collection	125
4. Temperature and salinity measurements and pink salmon enumeration	125
5. Results	127
6. Discussion	127
7. Conclusion	133
Acknowledgments	133
References	133

Abstract

Salmon lice (*Lepeophtheirus salmonis*) is an obligate marine macroparasite commonly found on Pacific salmonids. Due to its stenohaline restriction, salmon lice typically die shortly after adult salmon enter freshwater to spawn. During a study of juvenile salmonids diet in streams of different age after deglaciation within Glacier Bay in southeast Alaska, several gut samples contained salmon lice. The abundance of salmon lice in the diet of juvenile coho salmon (*Oncorhynchus kisutch*) and juvenile Dolly Varden (*Salvelinus malma*) was significantly correlated ($P < 0.05$) with salinity in the surface waters at 1 m in Glacier Bay. Salmon lice were absent in the diet of juvenile salmonids in younger streams towards the head of the Bay. Abundance of pink salmon (*Oncorhyncus gorbuscha*) adults which typically return in larger numbers than other species particularly in odd years, seemed the most important vector for salmon lice reaching freshwater systems. To our knowledge this is the first-time salmon lice have been documented in the diets of stream-dwelling juvenile salmonids, suggesting another route of marine derived N, P and C into freshwater food webs.

1. Introduction

Salmon lice (*Lepeophtheirus salmonis*), are a marine ectoparasitic copepod commonly found in the Northern Hemisphere (Connors et al., 2008a), one of several *Lepeophtheirus* species. *L. salmonis* will be referred to throughout this paper as salmon lice. While typically found on salmonids, several other non-salmonid hosts may occur (Jones et al., 2006; Kabata, 1979; Lyndon and Toovey, 2001; Nagasawa, 2004; Nekouei et al., 2018). Four Pacific salmonid species are found in all study streams in Glacier Bay watersheds, coho salmon, Dolly Varden, both rearing in the streams with varied life cycle adaptations, and pink and chum (*Oncorhynchus keta*) which go directly to the ocean after emerging from the gravel. Sockeye salmon (*Oncorhynchus nerka*) are found in three of the watersheds studied and rainbow trout (*Oncorhynchus mykiss*) in one stream. King salmon (*Oncorhynchus tshawytscha*) has so far not been found to spawn in Glacier Bay. Other species captured in several of the studied streams were three-spined stickleback (*Gasterosteus aculeatus*) and Coastrange sculpin (*Cottus aleuticus*).

Salmon lice were discovered in the diet of stream rearing juvenile salmonids in older streams during a dietary study within Glacier Bay in southeast Alaska. To understand why salmon lice was only found in older streams, we need to understand salmon lice life cycle and host interactions. Information around infections of juvenile Pacific salmon in areas without salmon farms is lacking, but key to understand the natural ecology of lice and salmon (Krkoek et al., 2007). The extent of infestation varies between species of fish, life stage, habitats and temporally in marine waters (Trudel et al., 2007; Wertheimer et al., 2003). Wertheimer et al. (2003) observed difference in infection of salmon lice between species of first summer at sea juvenile salmon, where prevalence increased for juvenile pink, chum and sockeye salmonids from June to July, then declined from July to August, except for coho salmon where it increased into late August. Jones et al. (2006) also found variation in development on three-spined stickleback, juvenile chum salmon and pink salmon in the coastal waters, but with mean abundance of salmon lice significantly higher on three-spined sticklebacks. In other studies, prevalence and abundance increased with host size and ocean age (Nagasawa, 1987), with pink and chum salmon accounting for 87% of the salmon lice population from 1991 to 1997 (Nagasawa, 2001). Gottesfeld et al. (2009) found a sharp increase in salmon lice abundance on juvenile pink salmon smolt near adult capture sites between May and July 2006 on the north cost of British Columbia. Wertheimer et al. (2003) observed salmon lice on first summer at sea salmonids in strait habitats southeast of GBNP, with prevalence of 4.6% for pink salmon, 10.4% for chum salmon, 26.9% for sockeye salmon and 52.7% for coho salmon in late July 2003. First summer at sea juvenile salmonids were less infected than adults of the same species, and low infestation rates were observed in May and June (Wertheimer et al., 2003). Orsi et al. (2002) observed generally higher salmonid catches in July for all species and all stages, juvenile, immature and adult, while sometimes higher for coho, chum and chinook later in the year, and higher in June for sockeye in the northern region of southeast Alaska. However, adult catches were low compared to juveniles (Orsi et al., 2002).

Salmon lice have eight life stages; two nauplii stages are non-feeding planktonic larvae, one infective planktonic copepod stage, two chalimus stages embedded on the host skin and two mobile pre-adults stages and one mobile adult stage that can move freely over the host skin (Hamre et al., 2013). Mennerat et al. (2012) observed a negative correlation between salmon lice fecundity and fish growth and that area with skin damage decreased over time while increased with infestation intensity. Jones et al. (2006) found increased salmon size was associated with decreased lice abundance but increased lice development. Higher levels of host immunity response might slow development and reduce fecundity of salmon lice (Mennerat et al., 2012).

Environmental conditions, especially salinity and water temperature have been found in many studies to impact salmon lice biology. Development time and survival are strongly influenced by water temperature and salinity (Johnson and Albright, 1991). Szetey et al. (2021) found different stages of salmon lice larva exhibit different vertical response to light, temperature and salinity. When the overlying layer became more brackish both nauplii and copepodids increasingly avoided it, and aggregated just below the halocline if present (Crosbie et al., 2019). Tucker et al. (2001) found both water temperature and salinity impacted the settlement and survival of copepodids. Survival and host infectivity of salmon lice are greatly reduced by short-term exposure to low salinity (Bricknell et al., 2006), and Jones et al. (2006) found low salinity zones coincided with low abundance of salmon lice. Johnson and Albright (1991) found that early and infective stages were most sensitive to low salinity, and salmon lice copepodids survived for < 24 h at ≤10‰ salinity. Samsing et al. (2016) found a negative relationship between water temperature and development times, adult body size, and reproductive outputs, except at 3 °C where larvae failed to reach the infective stage. Orsi et al. (2002) found the surface (2 m) water temperature in the Icy Strait to be higher in the summer months (June–August) while the salinity was lower in the same period compared to both May and September. Connors et al. (2008a) observed no salmon lice survival after 108 h in freshwater and no significant variation in survival between sexes. While salmon lice survival in freshwater varies, survival has been reported up to 3 weeks on Arctic charr (*Salvelinus alpinus*) (Finstad et al., 1995). Salmonids may return to fresh water to restore compromised osmotic and ionic imbalance (Bjørn et al., 2001; Wagner et al., 2004) and delouse in the process (Bjørn et al., 2001; Finstad et al., 1995). Environmental conditions will therefore impact salmon lice abundance and prevalence. The natural abundance of salmon lice is not well known, as most research is related to aquaculture and salmon lice as their most significant pathogen (Costello, 2009). Salmon lice from salmon farming can cause negative impact on wild salmonids (Shephard and Gargan, 2021; Torrissen et al., 2013) and Bjørn et al. (2001) observed significantly higher salmon lice infections at locations with salmon farming. Natural background levels of salmon lice are therefore important to better understand the negative influence of salmon farms.

The overarching aim of this study was to examine the diet of juvenile coho salmon and Dolly Varden in relation to different stages of watershed development since deglaciation and the potential change in food sources in Glacier Bay. Abiotic and biotic factors regulating salmon lice distribution and abundance in coastal areas are poorly understood (Jones and Beamish, 2011). Aggregation of Pacific salmon in coastal areas prior to entering their natal stream to spawn (Beamish et al., 2005), could provide an opportunity to study the autecology of salmon lice (Gottesfeld et al., 2009). Glacier Bay is a unique natural laboratory to study colonization and succession of stream systems, and the fjord provide a strong salinity gradient where the natural abundance of salmon lice can be studied (Milner et al., 2011). Salmon are keystone species and constitute links between marine and terrestrial habitats (Orsi et al., 2002). We elucidate here the prevalence of salmon lice in the diet of stream-dwelling juvenile coho salmon and Dolly Varden, and the potential relation to the salinity gradient found from younger to older watersheds due to deglaciation.

2. Methods
2.1 Study area
The study was undertaken within Glacier Bay in southeast Alaska, consisting of a fjord with two major arms, which have experienced deglaciation at

different points in time since the Little Ice Age (see Milner this issue). During the deglaciation of Glacier Bay, watersheds and streams emerge with different age and complexity. Five stream systems were examined; Stonefly Creek (SFC), Wolf Point Creek (WPC), Ice Valley Stream (IVS), Berg Bay South Stream (BBS), and Rush Point Creek (RPC) ranging from 33 to 200 years (when the study was undertaken) since the stream mouth was under ice. Determination of stream age used satellite and aerial photos, historical data, journal articles and unpublished data as outlined in Milner et al. (2000). The study streams in Glacier Bay and the location of CDT (conductivity, temperature, depth) stations are found in Fig. 1.

3. Sample collection

Juvenile coho salmon and Dolly Varden char were captured by minnow traps (400 mm × 220 mm, with 6 mm mesh), baited with salmon eggs previously soaked in Beta Dyne solution for sterilization. Minnow traps fished for 1.5 h at each site Jul 28th–Aug 11th in 2009, 2010 and 2011. In 2011 salmon eggs were enclosed inside "Kinder egg" capsules with small holes to prevent juvenile salmonids consuming the eggs, and thus also count the eggs found in their stomachs, since juveniles could not access to the baited minnow trap eggs. Captured juvenile coho salmon and Dolly Varden were sedated with clove oil, their guts evacuated by gastro-evacuation method of Meehan and Miller (1978) and stored in 70%–80% ethanol, and fish returned to the stream. No mortality occurred. At the laboratory, stomach contents were examined under a stereomicroscope and clearly identifiable material counted.

4. Temperature and salinity measurements and pink salmon enumeration

Water temperature and salinity measurements of the fjord were undertaken by the U.S. National Park Service from 2009 to 2011 in middle of July at 1 m increments. Nine measuring stations were located from the upper fjord parallel to WPC (site 1) southwards to Icy Strait (site 9), which correlates best with our streams, and a period when Glacier Bay would be expected to contain migrating adult salmonids returning to spawn. MATLAB was used to create the salinity contours by depth, and distance from station 1 to 9. Practical Salinity Units (PSU) were measured at depths from 1 to 40 m

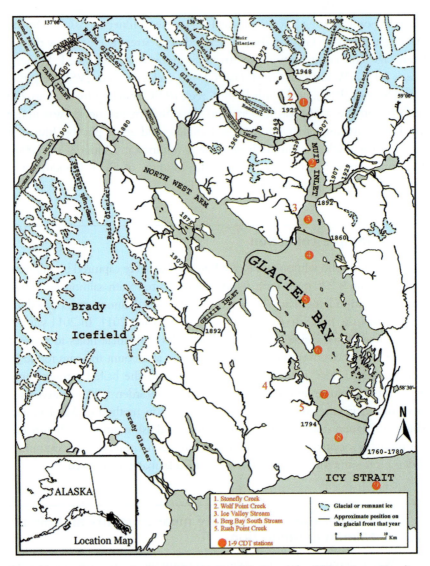

Fig. 1 The study streams in Glacier Bay National Park and the CDT stations. Map lines delineate study areas and do not necessarily depict accepted national boundaries.

in GBNP as most Pacific salmonids utilize these ocean depths (Morita, 2011; Walker et al., 2007). Spearman-rank correlation between average number of salmon lice (2009–2011) and average salinity (2009–2011) was undertaken in R Statistical software (2024.4.1.748) (R Core Team, 2024). Pink salmon spawner abundance was estimated by averaging counts of two observers walking the length of the stream during August.

5. Results

Salmon lice were found in the diet of juvenile coho salmon and Dolly Varden in the older Glacier Bay streams BBS and RPC, but not found in the younger upper bay streams, IVS, WPC and SFC (Table 1). Only female salmon lice (easily observed since all possessed egg strings) were observed in the diet, with significantly higher abundance found in juvenile coho and Dolly Varden diets in 2009 and 2011 at RPC and BBS, since only two salmon lice were found in RPC in 2010. In RPC 2011 both juvenile coho salmon and Dolly Varden consumed more salmon lice than salmon eggs. Although only around 150 adult pink salmon were counted on August 8th 2011 in RPC, 50 salmon lice were found in the guts of seven juvenile coho (17 salmon lice v 3 salmon eggs) and eight juvenile Dolly Varden (33 salmon lice v 29 salmon eggs). Total numbers of salmon lice were low compared to other prey items (although salmon lice were the main prey item in several juvenile salmonids), but higher than expected with so few spawning salmonids when the study was undertaken at RPC in 2011. Over 1100 adult pink salmon were enumerated in BBS during the same sampling period.

Mean salmon lice (n) per stream (2009–2011) in the diet (Fig. 2) correlated significantly (Spearman-rank correlation coefficient = 0.89) with mean salinity at 1 m (2009–2011) in the middle of July when the pink salmon spawners are prevalent throughout Glacier Bay. Glacier Bay has an increasing salinity gradient from the upper fjords to the mouth of the Bay, mainly due to freshwater input from glacial runoff. Salinity and water temperature contours are displayed in Fig. 3 with distance in km from CDT stations. Salinity increased southwards varying at 1 m from 12.5 PSU to 30.9 PSU from 2009 to 2011 from upper bay outside WPC to Icy Strait, with higher gradient in 2009 and 2011. CDT station number 2 was the closest to SFC (the youngest stream since deglaciation) and showed the lowest average salinity (2009–2011) at 1 m with 15.9 PSU. Glacier Bay has a strong tidal activity, but larger amount of freshwater is present in the upper east arm close to glacial influence. Surface water temperature was higher while salinity was lower in 2010 compared to 2009 and 2011.

6. Discussion

Salmon lice were not found in streams where the salinities proximal to the stream mouth were < 25 PSU, indicating adult salmon are free of

Table 1 Salmon lice prevalence (% fish that ate salmon lice) and range (variation in number of salmon lice found in one gut) (2009–2011) in all five streams, in addition to number of juveniles evacuated for diet content, number of empty guts and adult pink spawners.

Stream	Salmonid	Year 2009 % Prevalence (Range)	Juvenile (empty)	Adult Pink salmon	2010 % Prevalence (Range)	Juvenile (empty)	Adult Pink salmon	2011 % Prevalence (Range)	Juvenile (empty)	Adult Pink salmon
SFC	coho	0	30 (0)	15	0	30 (0)	1400	0	30 (0)	823
	Dolly Varden	0	29 (3)		0	30 (2)		0	30 (3)	
WPC	coho	0	28 (0)	950	0	26 (0)	9500	0	30 (2)	14130
	Dolly Varden	0	23 (0)		0	2 (0)		0	21 (1)	
IVS	coho	0	30 (0)	0	0	32 (0)	0	0	30 (2)	865
	Dolly Varden	0	24 (2)		0	2 (0)		0	30 (2)	
BBS	coho	30.00% (0–4)	30 (0)	2000	0	24 (0)	0	10.00% (0–1)	30 (3)	2675
	Dolly Varden	60.00% (0–3)	5 (1)		0	13 (3)		16.67% (0–3)	30 (3)	
RPC	coho	3.33% (0–1)	30 (1)	1000	0	30 (0)	209	22.58% (0–5)	31 (1)	1031
	Dolly Varden	3.45% (0–1)	29 (3)		6.67% (0–1)	30 (0)		26.67% (0–11)	30 (3)	

Salmon lice as a food source

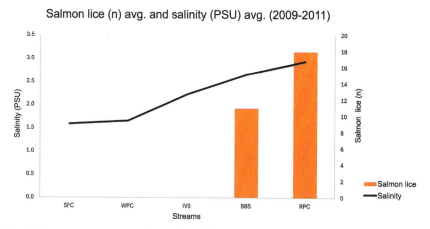

Fig. 2 Three-year average salmon lice (n) found in the diet and a three-year salinity (PSU) average.

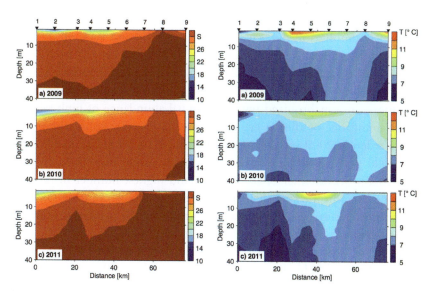

Fig. 3 (A–C) Salinity first column and temperature in second column for the first 40 m of depth at CDT station 1–9 (marked on top of column, with distance on the bottom).

salmon lice before entering natal freshwater streams. The increasing salinity gradient from the upper to lower bay streams therefore likely accounts for the difference in salmon lice occurrence in the diet of juvenile salmonids. The capability of salmon lice to remain attached to adult salmon is reduced in the upper bay area due to lower salinities and temperatures in the surface water at 1 m due to glacial runoff. The first form of defense against parasite infection is believed to be altered behavior (Bui et al., 2018), and water temperature and salinity influence salmon lice processes (Samsing et al., 2016; Tucker et al., 2001). Tanaka et al. (2000) found a positive correlation between 0 and 1 m surface water swimming adult chum salmon and colder surface water, which again may lead to a higher loss of salmon lice due to prolonged time in colder and low salinity water. Furevik et al. (1993) observed greater jumping activity with increased lice infestation on Atlantic salmon (*Salmo salar*) in salmon farm net pens. Loss of salmon lice due to jumping in Glacier Bay would likely be enhanced by the substantially lower salinity and temperature in the surface water. Jones et al. (2006) observed a spatial relationship between salmon lice abundance and the salinity of the surface water. Surface water salinity further south between IVS and BBS streams markedly changed, with waters proximal to BBS and RPC above 25 PSU. Lower surface water temperature in 2010, in addition to lower salinity, may have reduced the influence of salmon lice in the diets of juvenile salmonids in the lower bay streams BBS and RPC in 2010.

Adult pink salmon spawners are probably the most important host for salmon lice (Nagasawa, 1987, 2001; Nagasawa and Takami, 1993). Nagasawa (2001) found that the overall annual salmon lice population fluctuations are synchronized with the annual abundance of pink salmon. Higher abundance of salmon lice in the diet of juvenile coho salmon and Dolly Varden in 2009 and 2011 compared to 2010 could be linked to the higher odd year run of pink salmon in northern southeast Alaska (Milner et al., 2008). Wertheimer et al. (2003) found highest number of infected juvenile salmon in the strait habitats in late July, which coincide with the pink salmon return. Returning adult salmon under natural conditions seems therefore to be the most important source of salmon lice (Gottesfeld et al., 2009). Salmon lice population decreased in strait habitats from August, which could be due to first summer at sea juveniles continue their outmigration away from these habitats (Wertheimer et al., 2003). Salmon lice on later returning salmon may therefore stay attached potentially longer due to fewer suitable hosts in Icy Strait and Glacier Bay. Fish gut samples were collected at the beginning of the upstream pink salmon runs in early August, and higher salmon lice content in the diet might therefore be expected at the peak of the instream adult pink salmon run. Chum salmon entered our study streams a few days later than pink salmon and coho salmon run later in the year, and as salinity gradually increases in the 2 m surface water throughout September (Orsi et al., 2002), could additionally lead to a higher number of salmon lice still attached when adult salmon reach their natal stream. However, during the same time the water temperature decreased (Orsi et al., 2002) which could lead to a decrease in numbers of salmon lice still attached. Timings of salmon runs vary from year to year, often regulated by stream water level, as well as sampling time, which could contribute to variability in these findings. Stream water level at IVS was on several occasions extremely low or even absent from the stream bed. The peak of pink salmon runs, and other later spawning Pacific salmon could increase the salmon lice available to resident juvenile salmonids. However, due to low abundance of other species of salmon compared to adult pink salmon (Milner et al., 2008), their salmon lice contribution to the diet is expected to be low.

The route by which salmon lice become incorporated into the diets of juvenile salmonids is unknown. Mortality may occur or detachment from the adult salmon and then consumed from the drift or directly removed by the juvenile salmonids from adult fish. Experimental studies have found that salmon lice tend to become detached from their host over time (Finstad et al., 1995), and salmon lice missing on juvenile pink and chum salmon in holding tubs were expected to be eaten by the juveniles after dying and falling off their host (Connors et al., 2008a). During an experiment Morton and Routledge (2005) observed that the juvenile pink and chum salmon picked adult salmon lice off each other. Losos et al. (2010) found cleaning behavior (picking salmon lice off pink juvenile salmon) in sticklebacks during an experiment in the Broughton Archipelago, British Colombia, Canada. Due to salmon lice mortality when entering freshwater, possibly with help of abrasion during crowding of adult salmon going up-stream, diet contribution through drift seems the most probable route, since both juvenile coho and Dolly Varden are more or less drift feeders (Kaeriyama and Nakano, 1995).

Only adult female salmon lice were found in the diet of juvenile coho salmon and Dolly Varden, perhaps due to difference in behavior and morphology constraining escape (Connors et al., 2008b) from adult salmon entering natal streams. Hull (1998) observed that the mean adult male number inter-host transferrers were approximately 3.5 times greater than in mated females, and that transferring males tended to transfer more often. Salmon lice also escape predation on their hosts by moving from prey to predator, and male salmon lice were found to transfer 3.8 times more frequently than females (Connors et al., 2008b). Male salmon lice can mate several times and make no further investment in the fertilized progeny

(Hull, 1998). Generally, the higher catch of most salmonids occurs in July and August except for coho, and higher loss of salmon lice could be expected before salmon enter their natal stream. Both the temperature and salinity are higher and lower respectively during June–August than May and September (Orsi et al., 2002) Tens of millions adult pink and chum salmon return through Icy Strait in June and July, and the higher infestation rates in July may reflect concentration of salmon lice-bearing adult salmon passing through (Wertheimer et al., 2003). Therefore, male salmon lice could be expected to be already detached when adult salmon enter freshwater. Interannual variation of salmon lice could also impact the number of salmon lice still attached to their host when adult salmon return. Previous studies (e.g. Wertheimer et al., 2003) potentially indicate that the window where salmon lice could end up in the diet of stream dwelling salmonids in Glacier Bay is narrow, due to the high numbers of salmon lice found in Chatham Strait and Icy Strait (southeast of Glacier Bay) in late July and early Augst, compared to late June and late August.

When salmon eggs were enclosed in "Kinder egg" capsules in 2011 female salmon lice were more common in the guts than salmon eggs in RPC for juvenile coho and Dolly Varden (22.58% vs 3.23%, 26.67% vs 23.33% respectively) as a comparison could be made that year. Johnson and Ringler (1979) showed that Pacific salmon eggs accounted for at least 90% of the October diet of stream-dwelling juvenile salmonids and appeared responsible for a large increase in condition factor. While attached salmon lice could last for more than a week in fresh water, mortality will occur after 8 h of free swimming (Hahnenkamp & Fyhn, 1985). The time adult salmon spent in the Bay could be important in relation to the number of salmon lice that still would be attached when adult salmon enter their natal streams and become available to juvenile coho salmon and Dolly Varden. Salmon returning to spawn have an important role in the contribution of marine derived nutrients (MDN) to their natal streams as they accumulate 95% of their biomass in the ocean (Schindler et al., 2003). Stable isotope analyses of adult female salmon lice showed high $\delta^{15}N$ and $\delta^{13}C$ levels, with all samples from salmon lice above > 12‰ and > −20‰ respectively (Dean et al., 2011). Salmon lice therefore also contribute MDN in addition to the potential nutrient value due to their physically large size, compared to other in-stream diet items. While salmon lice will be available only for a short time, it could still play a role in the diet of stream-dwelling juvenile coho salmon and Dolly Varden. Further studies would be needed to see if the MDN could be utilized effectively by stream-dwelling juvenile coho salmon and Dolly Varden.

7. Conclusion

To our knowledge this is the first time that salmon lice have been documented as a food source for stream-dwelling juvenile salmonids. Salinity in estuarine and proximal waters to watersheds seems to play an important role in determining whether salmon lice are found in freshwater. The spatial difference from younger to older streams provide insight into the salmon lice autecology in the Glacier Bay. This is another route by which MDN can become incorporated into stream food webs and serve as an additional food source for juvenile salmonids. However, we do not know the actual nutritional value of salmon lice or the abundance of salmon lice that would still be attached adult salmon after entering their natal streams. More research should therefore be carried out to fill in these knowledge gaps. Accurate estimates of salmon lice natural autecology, especially hosts interactions are essential to understand abiotic and biotic processes that affects their natural productivity. The possible synergism of meltwater and glacial runoff causing low salinity and lower surface water temperature, and the absence of salmon farms in the vicinity, make Glacier Bay a perfect place to study the natural background and baseline levels of salmon lice in fjord systems.

Acknowledgments

We thank Edward Malone, Megan Klaar, Debra Finn and Laura German for assistance in the field. Thanks to the National Park Service for the salinity data and for logistical support to reach the study streams, especially Justin Smith Captain of the mv Capelin. We are very grateful to Ilker Fer at Bergen University for assistance with MATLAB.

References

Beamish, R.J., Neville, C.M., Sweeting, R.M., Ambers, N., 2005. Sea lice on adult Pacific salmon in the coastal waters of Central British Columbia, Canada. Fish. Res. 76, 198–208. https://doi.org/10.1016/j.fishres.2005.06.007.

Bjørn, P.A., Finstad, B., Kristoffersen, R., 2001. Salmon lice infection of wild sea trout and Arctic char in marine and freshwaters: the effects of salmon farms. Aquac. Res. 32, 947–962. https://doi.org/10.1046/j.1365-2109.2001.00627.x.

Bricknell, I.R., Dalesman, S.J., O'Shea, B., Pert, C.C., Luntz, A.J.M., 2006. Effect of environmental salinity on sea lice *Lepeophtheirus salmonis* settlement success. Dis. Aquat. Organ. 71, 201–212. https://doi.org/10.3354/dao071201.

Bui, S., Oppedal, F., Samsing, F., Dempster, T., 2018. Behaviour in Atlantic salmon confers protection against an ectoparasite. J. Zool. 304, 73–80. https://doi.org/10.1111/jzo.12498.

Connors, B.M., Juarez-Colunga, E., Dill, L.M., 2008a. Effects of varying salinities on *Lepeophtheirus salmonis* survival on juvenile pink and chum salmon. J. Fish. Biol. 72, 1825–1830. https://doi.org/10.1111/j.1095-8649.2008.01839.x.

Connors, B.M., Krkošek, M., Dill, L.M., 2008b. Sea lice escape predation on their host. Biol. Lett. 4, 455–457. https://doi.org/10.1098/rsbl.2008.0276.

Costello, M.J., 2009. How sea lice from salmon farms may cause wild salmonid declines in Europe and North America and be a threat to fishes elsewhere. Proc. R. Soc. Lond. B: Biol. Sci. 276, 3385–3394. https://doi.org/10.1098/rspb.2009.0771.

Crosbie, T., Dw, W., Oppedal, F., IA, J., Samsing, F., Dempster, T., 2019. Effects of step salinity gradients on salmon lice larvae behaviour and dispersal. Aquac. Environ. Interact. 11. https://doi.org/10.3354/aei00303.

Dean, S., DiBacco, C., McKinley, R.S., 2011. Assessment of stable isotopic signatures as a means to track the exchange of sea lice (*Lepeophtheirus salmonis*) between host fish populations. Can. J. Fish. Aquat. Sci. 68, 1243–1251. https://doi.org/10.1139/f2011-039.

Finstad, B., Bjørn, P.A., Nilsen, S., 1995. Survival of salmon lice, *Lepeophtheirus salmonis* Krøyer, on Arctic charr, *Salvelinus alpinus* (L.), in fresh water. Aquac. Res. 26, 791–795. https://doi.org/10.1111/j.1365-2109.1995.tb00871.x.

Furevik, D.M., Bjordal, Å., Huse, I., Fernö, A., 1993. Surface activity of Atlantic salmon (*Salmo salar* L.) in net pens. Aquaculture 110, 119–128. https://doi.org/10.1016/0044-8486(93)90266-2.

Gottesfeld, A.S., Proctor, B., Rolston, L.D., Carr-Harris, C., 2009. Sea lice, *Lepeophtheirus salmonis*, transfer between wild sympatric adult and juvenile salmon on the north coast of British Columbia, Canada. J. Fish. Dis. 32, 45–57. https://doi.org/10.1111/j.1365-2761.2008.01003.x.

Hahnenkamp, L., Fyhn, H., 1985. The osmotic response of salmon louse, *Lepeophtheirus salmonis* (Copepoda: Caligidae), during the transition from sea water to fresh water. J. Comp. Physiol. B 155, 357–365. https://doi.org/10.1007/BF00687479.

Hamre, L.A., Eichner, C., Caipang, C.M.A., Dalvin, S.T., Bron, J.E., Nilsen, F., et al., 2013. The salmon louse *Lepeophtheirus salmonis* (Copepoda: Caligidae) life cycle has only two chalimus stages. PLoS ONE 8, e73539. https://doi.org/10.1371/journal.pone.0073539.

Hull, M.Q., 1998. Patterns of pair formation and mating in an ectoparasitic caligid copepod *Lepeophtheirus salmonis* (Krøyer 1837): implications for its sensory and mating biology. Philos. Trans. R. Soc. B: Biol. Sci. 353, 753–764. https://doi.org/10.1098/rstb.1998.0241.

Johnson, S.C., Albright, L.J., 1991. Development, growth, and survival of *Lepeophtheirus salmonis* (Copepoda: Caligidae) under laboratory conditions. J. Mar. Biol. Assoc. U. K. 71, 425–436. https://doi.org/10.1017/S0025315400051687.

Johnson, J.H., Ringler, N.H., 1979. Predation on pacific salmon eggs by salmonids in a tributary of Lake Ontario. J. Gt. Lakes Res. 5, 177–181. https://doi.org/10.1016/S0380-1330(79)72144-7.

Jones, S., Beamish, R.J., 2011. Salmon Lice: An Integrated Approach to Understanding Parasite Abundance And Distribution. https://doi.org/10.1002/9780470961568.

Jones, S.R.M., Prosperi-Porta, G., Kim, E., Callow, P., Hargreaves, N.B., 2006. The occurrence of *Lepeophtheirus salmonis* and *Caligus demensi* (Copepoda: Caligidae) on three-spine stickleback gasterosteus aculeatus in Coastal British Columbia. J. Parasitol. 92, 473–480.

Kabata, Z. 1979. Parasitic Copepoda of British Fishes. Ray Society.

Kaeriyama, M., Nakano, S., 1995. Summer microhabitat use and diet of four sympatric stream-dwelling salmonids in a Kamchatkan stream. Fish. Sci. 61, 926–930. https://doi.org/10.2331/fishsci.61.926.

Krkoek, M., Gottesfeld, A., Proctor, B., Rolston, D., Carr-Harris, C., Lewis, M., 2007. Effects of host migration, diversity and aquaculture on sea lice threats to Pacific salmon populations. Proc. Biol. Sci./R. Soc. 274, 3141–3149. https://doi.org/10.1098/rspb.2007.1122.

Losos, C., Reynolds, J., Dill, L., 2010. Sex-selective predation by threespine sticklebacks on sea lice: a novel cleaning behaviour. Ethology 116, 981–989. https://doi.org/10.1111/j.1439-0310.2010.01814.x.

Lyndon, A., Toovey, J.P.G., 2001. Occurrence of gravid salmon lice (*Lepeophtheirus salmonis* (Krøyer)) on saithe (*Pollachius virens* (L.)) from salmon farm cages. Bull. Eur. Assoc. Fish. Pathol. 21, 84–85.

Meehan, W.R., Miller, R.A., 1978. Stomach flushing: effectiveness and influence on survival and condition of juvenile salmonids. J. Fish. Res. Bd. Can. 35, 1359–1363. https://doi.org/10.1139/f78-212.

Mennerat, A., Hamre, L., Ebert, D., Nilsen, F., Dávidová, M., Skorping, A., 2012. Life history and virulence are linked in the ectoparasitic salmon louse *Lepeophtheirus salmonis*. J. Evolut. Biol. 25, 856–861. https://doi.org/10.1111/j.1420-9101.2012.02474.x.

Milner, A.M., Knudsen, E.E., Soiseth, C., Robertson, A.L., Schell, D., Phillips, I.T., et al., 2000. Colonization and development of stream communities across a 200-year gradient in Glacier Bay National Park, Alaska, USA. Can. J. Fish. Aquat. Sci. 57, 2319–2335.

Milner, A.M., Robertson, A.L., Brown, L.E., Sønderland, S.H., McDermott, M., Veal, A.J., 2011. Evolution of a stream ecosystem in recently deglaciated terrain. Ecology 92, 1924–1935.

Milner, A., Robertson, A., Monaghan, K., Veal, A., Flory, E., 2008. Colonization and development of an Alaskan stream community over 28 years. Front. Ecol. Environ. 6, 413–419. https://doi.org/10.1890/060149.

Morita, K., 2011. Body size trends along vertical and thermal gradients of chum salmon in the Bering Sea during summer. Fish. Oceanogr. 20, 258–262.

Morton, A., Routledge, R., 2005. Mortality rates for juvenile pink *Oncorhynchus gorbuscha* and Chum O. keta Salmon infested with sea lice lepeophtheirus salmonis in the broughton archipelago. Alsk. Fish. Res. Bull. 11, 146–152.

Nagasawa, K., 1987. Prevalence and abundance of *Lepeophtheirus salmonis* (Copepoda: Caligidae) on high-seas salmon and trout in the North Pacific Ocean. Nippon. Suisan Gakkaishi 53, 2151–2156. https://doi.org/10.2331/suisan.53.2151.

Nagasawa, K., 2001. Annual changes in the population size of the salmon louse Lepeophtheirus salmonis (Copepoda: Caligidae) on high-seas Pacific Salmon (*Oncorhynchus* spp.), and relationship to host abundance. In: Lopes, R.M., Reid, J.W., Rocha, C.E.F. (Eds.), Copepoda: Developments in Ecology, Biology and Systematics, Developments in Hydrobiology. Springer Netherlands, pp. 411–416.

Nagasawa, K., 2004. Sea lice, *Lepeophtheirus salmonis* and *Caligus orientalis* (Copepoda: Caligidae), of wild and farmed fish in sea and brackish waters of Japan and adjacent regions: a review. Zool. Stud. 43, 173–178.

Nagasawa, K., Takami, T., 1993. Host utilization by the Salmon Louse *Lepeophtheirus salmonis* (Copepoda: Caligidae) in the Sea of Japan. J. Parasitol. 79, 127–130. https://doi.org/10.2307/3283292.

Nekouei, O., Vanderstichel, R., Thakur, K., Arriagada, G., Patanasatienkul, T., Whittaker, P., et al., 2018. Association between sea lice (*Lepeophtheirus salmonis*) infestation on Atlantic salmon farms and wild Pacific salmon in Muchalat Inlet, Canada. Sci. Rep. 8. https://doi.org/10.1038/s41598-018-22458-8.

Orsi, J.A., Fergusson, Emily, A., Heard, William, R., Mortensen, D.G., et al., 2002. Survey of juvenile salmon in the marine waters of southeastern Alaska, May–September 2001. NPAFC Doc. 630 99801–8626.

R Core Team, 2024. R: A Language and Environment for Statistical Computing. R Foundation for Statistical Computing, URL https://www.R-project.org/. Vienna, Austria.

Samsing, F., Oppedal, F., Dalvin, S., Johnsen, I., Vågseth, T., Dempster, T., 2016. Salmon lice (*Lepeophtheirus salmonis*) development times, body size and reproductive outputs follow universal models of temperature dependence. Can. J. Fish. Aquat. Sci. 73. https://doi.org/10.1139/cjfas-2016-0050.

Schindler, D.E., Scheuerell, M.D., Moore, J.W., Gende, S.M., Francis, T.B., Palen, W.J., 2003. Pacific salmon and the ecology of coastal ecosystems. Front. Ecol. Environ. 1, 31–37. https://doi.org/10.1890/1540-9295(2003)001[0031:PSATEO]2.0.CO;2.

Shephard, S., Gargan, P., 2021. Wild Atlantic salmon exposed to sea lice from aquaculture show reduced marine survival and modified response to ocean climate. ICES J. Mar. Sci. 78, 368–376. https://doi.org/10.1093/icesjms/fsaa079.

Szetey, A., Wright, D., Oppedal, F., Dempster, T., 2021. Salmon lice nauplii and copepodids display different vertical migration patterns in response to light. Aquac. Environ. Interact. 13. https://doi.org/10.3354/aei00396.

Tanaka, H., Takagi, Y., Naito, Y., 2000. Behavioural thermoregulation of chum salmon during homing migration in coastal waters. J. Exp. Biol. 203, 1825–1833.

Torrissen, O., Jones, S., Guttormsen, A., Asche, F., Horsberg, T., Skilbrei, O., et al., 2013. Sea lice—impacts on wild salmonids and salmon aquaculture. J. Fish. Dis. 36, 171–194.

Trudel, M., Jones, S.R., Thiess, M.E., Morris, J.F., Welch, D.W., Sweeting, R.M., et al., 2007. Infestations of motile salmon lice on Pacific salmon along the west coast of North America. In: American Fisheries Society Symposium. American Fisheries Society, p. 157.

Tucker, C., Sommerville, C., Wootten, R., 2001. The effect of temperature and salinity on the settlement and survival of copepodids of *Lepeophtheirus salmonis* (Krøyer, 1837) on Atlantic salmon, *Salmo salar* L. J. Fish. Dis. 23, 309–320. https://doi.org/10.1046/j.1365-2761.2000.00219.x.

Wagner, G.N., Mckinley, R.S., Bjørn, P.A., Finstad, B., 2004. Short-term freshwater exposure benefits sea lice-infected Atlantic salmon. J. Fish. Biol. 64, 1593–1604. https://doi.org/10.1111/j.0022-1112.2004.00414.x.

Walker, R.V., Sviridov, V.V., Urawa, S., Azumaya, T., 2007. Spatio-temporal variation in vertical distributions of Pacific salmon in the ocean. North. Pac. Anadromous Fish. Comm. Bull. 4, 193–201.

Wertheimer, A.C., Fergusson, Emily, A., Focht, R.L., Heard, W.R., Orsi, J.A., et al., 2003. Sea lice infection of juvenile salmon in the marine waters of the northern region of southeastern Alaska, May–August 2003 NPAFC Doc 706, 13.

CHAPTER SIX

Convergence of beta diversity in river macroinvertebrates following repeated summer floods

Lawrence J.B. Eagle[a,*], Alexander M. Milner[b,c], Megan J. Klaar[a], Jonathan L. Carrivick[a], and Lee E. Brown[a]

[a]School of Geography, University of Leeds, Woodhouse Lane, Leeds, United Kingdom
[b]School of Geography, Earth and Environmental Sciences, University of Birmingham, Edgbaston, Birmingham, United Kingdom
[c]Institute of Arctic Biology, University of Alaska Fairbanks, Fairbanks, AK, United States
*Corresponding author. e-mail address: lawrenceeaglejb@gmail.com

Contents

1. Introduction	138
2. Study area	141
3. Datasets and methods	144
4. Results	148
4.1 Community composition and summary metrics	148
5. Within site beta-diversity	151
6. Between river beta-diversity	153
7. Discussion	154
7.1 Short term impacts of the floods	154
7.2 Contrasting compositional response following the sequence of floods	156
7.3 Post-flood patterns of community change	157
8. Conclusion	159
Acknowledgements	160
Appendix 1	160
References	162

Abstract

Climate change is expected to increase in frequency, duration and magnitude of floods potentially driving substantial change in river ecosystems. Knowledge of flood effects on river biological communities is generally focused on individual high-magnitude floods. Here we demonstrate multiple effects of repeated atypical summer floods across three rivers in Glacier Bay, southeast Alaska at different stages of development following glacier retreat. Immediate short-term impacts (1st year post-flood) reduced mean invertebrate taxonomic richness from 22.3 (95% CI: 17.7, 27.0) to 13.6 (95% CI: 7.0, 20.3) and density from 9045 (ind/m^2, 95% CI: 5431,14765) to 1636 (95% CI: 812, 3641) across the three rivers. Recovery of these richness and density metrics occurred over two subsequent years of post-flood community response.

Persistent shifts in community composition were recorded in all rivers, with a decline in Jaccard's dissimilarity within each site from pre-flood 0.40 (95% CI: 0.25, 0.56) to 0.19 (95% CI: 0.03, 0.35) and between the three rivers from pre-flood 0.64 (95% CI: 0.56, 0.71) to 0.45 (95% CI: 0.38, 0.51) during the post-flood response. Convergence of community composition was associated with the predominant colonisation/re-establishment of chironomids and some small ephemeropterans in all rivers post-flood. Overall, these findings lead us to suggest that the flood sequences, including repeated atypical summer floods, can influence river biological community reassembly by favouring a core group of disturbance tolerant taxa.

1. Introduction

Floods are a defining component of river flow regimes (Poff et al., 1997). High flow disturbances are increasing in frequency and magnitude globally under climate change (Berghuijs et al., 2017) driven primarily by increased magnitude, frequency and duration of precipitation events (Martel et al., 2021; Trenberth, 2011; Witze, 2018). Such increases are observable in many regions of the world (Donat et al., 2016), with marked increases in North America over the last 50 years (Groisman et al., 2005). River responses to floods depend on their form (magnitude, frequency, duration, timing, and rate of change of hydrologic conditions) (Lytle & Poff, 2004; Poff et al., 1997). However, major questions remain concerning how changes in these components of flood disturbances will modify river ecosystems with the majority of research focussed on individual high magnitude floods often in single river basins.

Floods occur with seasonal predictability in many regions (Black & Werritty, 1997; Urrea et al., 2019; Yen & Chen, 2000), which is critical for enabling the completion of life cycles and providing an influx of nutrients from headwaters and floodplains (Junk et al., 1989). These seasonal events can be important for the persistence of river biological communities, acting as environmental filters of community composition (Poff, 1997). However, there is increasing evidence for changes in the timing and frequency of large floods across continental scales (Mallakpour & Villarini, 2015), with increased effects at higher latitudes (Blöschl et al., 2017). Furthermore, occurrence of major rainfall is less predictable outside of dominant seasonal wet periods, such as during summer in North America (Gianotti et al., 2013). The trend for increasing magnitude of floods has been linked to an increasing spatial extent of these events (Kemter et al., 2020). As changes to the seasonal timing of larger and more frequent floods occurs there is the potential for alterations to river ecosystems that have evolved under predictable rainfall-runoff regimes, however, this has not received major consideration in the literature to date.

Changes to flood regimes (increasing frequency and magnitude) influence the occurrence of extreme floods, characterised by peak discharge recurrence intervals > 50–100 years (Woodward et al., 2015). Extreme floods can drive persistent changes to channel geomorphology (Guan et al., 2015; Pasternack & Wyrick, 2017; Staines & Carrivick, 2015; Carrivick & Tweed, 2019; Tomczyk et al., 2020) and can substantially impact biotic elements of ecosystems (Milner et al., 2013; Woodward et al., 2015). However the extent of biological impacts can vary substantially between individual floods and taxonomic groups (Glover et al., 2020; Poff et al., 2018; Robertson et al., 2015). The form of a disturbance likely influences the extent to which it will impact upon an ecosystem (Radchuk et al., 2019). Floods with diverse forms (including atypical timings) have the potential to drive physical and biological change in river ecosystems. Few studies have addressed a series of atypically timed or formed flood disturbances.

Sequences of large floods (Guan et al., 2016) as well as smaller floods and high flow events (Bertoldi et al., 2010), here referred to as repeated floods, have been shown to drive substantial morphological change in some rivers (Eagle et al., 2021; Kavan et al., 2017; Warburton, 1994). The ecological effects of repeated floods have predominantly been explored in anthropogenically altered ecosystems, particularly below dams, driving persistent changes in river communities (Gillespie et al., 2020; Robinson, 2012; Robinson et al., 2018; Robinson & Uehlinger, 2008; Uehlinger et al., 2003). However, such studies may not be representative of the response of "natural" ecosystems to repeated floods. Response of natural watershed ecosystems to repeated floods has only been explored by one study of one river within Glacier Bay National Park, Alaska, USA (Milner et al., 2018). Significantly, Milner et al. (2018) demonstrated the capacity of both extreme high-magnitude and repeated floods to shift community composition immediately post-flood. Patterns of multi-year response during community reassembly have yet to be explored following repeated floods. Yet this understanding is necessary to identify the potential implications of increasingly frequent floods under climate change.

The response of a riverine community to flood disturbance is influenced by the taxa which are present at the time of the disturbance (Townsend & Hildrew, 1994). Taxa within a pre-flood community can be resistant (Pimm, 1984) and/or be resilient to community change

(Chambers et al., 2019) or become extinct (Ledger et al., 2012) depending upon the traits of each species present and availability of suitable refugia (Chase, 2007). Here resilience is defined as incorporating two components of disturbance response (1) *resistance* the capacity of taxa to persist through a disturbance and/or (2) *recovery* the capacity of taxa to rapidly re-establish/ establish post-disturbance (Holling, 1973). The taxa that persist post-flood may influence community reassembly via legacy effects (Ledger et al., 2006), while rapid colonisers may influence subsequent assembly via priority effects (Fraser et al., 2015). Following disturbance taxa previously excluded, as a result of species interactions, may be able to colonise flood impacted communities from nearby refugia, such as upstream tributaries. Reassembly may also be driven by colonisation of taxa from surrounding areas via aerial dispersal (Brown et al., 2018; Leibold et al., 2004). Overall, a combination of these process may occur together to drive community reassembly following major disturbances. Currently few cases exist where responses have been studied across multiple sites and years thereby limiting our capacity to understand the potential role of these processes.

Disturbance studies have reported rapid recovery under varied disturbance regimes (McMullen et al., 2017; White et al., 2022). In high magnitude flood studies, rapid post-flood recovery of total density is a consistent trait of macroinvertebrate communities (Marino et al., 2024; Mundahl and Hunt, 2011; Sabater et al., 2023; Woodward et al., 2015), but changes in community composition have been reported post-flood (Milner et al., 2013). Indeed, changes in taxonomic composition (to varying degrees) are a routine feature of multi-year response to both extreme high-magnitude and mostly artificial (dam release) repeated floods (Milner et al., 2018; Mundahl & Hunt, 2011; Rader et al., 2008; Robertson et al., 2015). However there has been little focus on repeated precipitation driven floods that occur in natural systems. Chironomidae species have been reported as rapid post-flood colonisers becoming abundant in early post-flood communities and the group is often considered highly resilient to floods (Mundahl & Hunt, 2011; Robertson et al., 2015). In contrast, resistance of baetid mayflies has been shown to be low, with abundance strongly declining pre- to post-flood, yet the taxa demonstrate high resilience by rapidly rebounding during post-flood community reassembly (Mundahl & Hunt, 2011). Such an understanding of post-flood taxonomic response for repeated floods is currently missing from the literature.

A sequence of high-frequency lower-magnitude floods (repeated floods) occurred across southeast Alaska during an atypical time of the year (summer 2014) providing an opportunity to enhance our understanding of both the first year immediate impact and subsequent two year macroinvertebrate community response to repeated atypical summer floods. Benthic macroinvertebrates, a well-studied group of organisms with well-defined taxonomic traits and well understood environmental requirements, are used as a "model" to understand ecological responses to flooding in this study (Jacobsen et al., 2012). This study aimed to test three main hypotheses:

H_1: repeated atypical summer floods will be associated with altered community composition and decreases in taxa richness, Shannon's diversity and total density across all sites one year post-flood, due to physical change in river channels (Eagle et al., 2021) and removal of flood sensitive taxa (Herbst & Cooper, 2010).

H_2: repeated atypical summer floods will act as strong environmental filters and homogenisation of the benthic macroinvertebrate community composition will be observed post-flood, leading to a decrease in both within site and between site river beta-diversity following the floods;.

H_3: post-flood reassembly will be dominated by a shared group of taxa including Chironomidae (Anderson & Ferrington, 2013) and small Ephemeroptera (Herbst & Cooper, 2010). These groups include numerous regionally abundant taxa (Milner et al., 2000) that can disperse effectively and rapidly exploit opportunities in disturbed ecosystems (Poff et al., 2018).

2. Study area

Glacier Bay National Park and Preserve (GBNP) (58°10′–59°15′ N.; 135°15′–138°10′ W.; Fig. 1) spans an area of 11,030 km² in SE Alaska, which was mostly covered by a Neoglacial ice sheet as recently as 1700. Subsequent glacier retreat exposed a large tidal fjord, 150 km long and 20 km wide, and surrounding terrestrial landscapes (Field, 1947; Hall et al., 1995) These recently exposed landscapes became available for colonisation through primary terrestrial (Chapin et al., 1994; Fastie, 1995) and aquatic succession (Milner et al., 2007).

In GBNP long-term river ecological sampling has been in place since the late 1970s (Milner, 1987; Milner & Bailey, 1989). The temporal development of freshwater environments results in a spatial gradient of physical

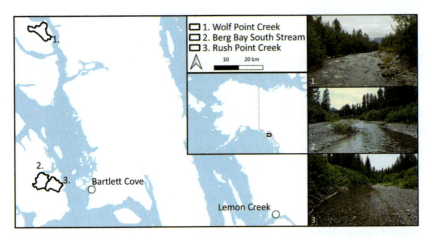

Fig. 1 Glacier Bay map, including the location of study river catchments (black), Bartlett Cove long term precipitation gauge and Lemon Creek where a long-term average daily discharge dataset is available. Images of river channels proximal to biological sampling locations on each study rivers.

environments and ecological communities (Klaar et al., 2009; Milner et al., 2000). Catchments of older rivers closer to the mouth of Glacier Bay have extensive forest cover, including along parts of river banks, while those of younger rivers possess catchments dominated by alder and willow scrub combined with some cottonwood (Klaar et al., 2015). While the effect of differences in successional stage between watersheds on river response to floods is not the focus of this study given the lack of river systems replicated by age, two of the three study rivers were older (RPC—209–214 years and BBS—184–189 years) and one younger (WPC—68–73 years). As such there is some potential for differences in response to floods given differences in pre-flood communities and physical environments linked to long-term successional differences (Klaar et al., 2009, 2015; Milner et al., 2000).

Glacier Bay experienced the wettest summer in a 31 year gauge precipitation record at Bartlett Cove in 2014 (Fig. 2, Appendix 1), averaging 189 mm of precipitation per month from June to August in contrast to the 31 year mean of 110 mm per month (Menne et al., 2012). Notable precipitation events continued until mid-October. During summer 2014 three of the ten wettest summer weeks on record occurred, with >100 mm falling in each week. The second wettest June and July on record were followed by a two-week period, from the 10th August, where approximately 200 mm of rain fell leading to repeated floods (Milner et al., 2018). Similar precipitation levels were observed at sites across GBNP in radar

River invertebrates and floods

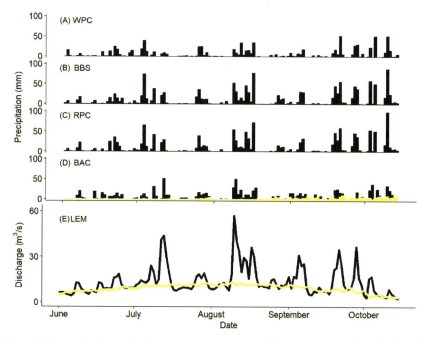

Fig. 2 Daily precipitation totals (mm) at (A) Wolf Point Creek (WPC), (B) Berg Bay South Stream (BBS), (C) Rush Point Creek, (D) Bartlett Cove (BAC), Glacier Bay (black), overlain with median daily precipitation total at Bartlett Cove, Glacier Bay (yellow) over 31 years (1987–2017). (E) Lemon Creek (LEM) daily mean discharge (black) hydrograph. Median daily mean discharge (yellow) over 33 years (1952–1973 & 2002–2016).

data and indeed across SE Alaska (Brettschneider, 2014; Huffman et al., 2019; Menne et al., 2012). Due to wilderness legislation, there are no long-term discharge datasets available in Glacier Bay. However, a long-term discharge record (33 years) was available for Lemon Creek, a proximal stream with detailed discharge data (serviced by USGS 80 km east near Juneau), eight notable high flow events occurred during the summer, with daily mean discharges regularly exceeding twice the daily median. These included three floods over nine days (10th August 2014–18th August 2014; Fig. 2), including the 7th highest daily mean discharge (10th August 2014) recorded in the time series (Eagle et al., 2021). The summer timing of these floods in quick succession make them atypical in the region where spring and autumn floods are the norm.

Importantly, the floods were geomorphologically effective in the three rivers included in this study which span the latitudinal range of the National Park. Greater longitudinal baseflow planform channel change occurred from pre- to post-flood than observed between two pre-flood surveys at all sites (Eagle et al., 2021). Repeat surveys of pairs of cross sections adjacent to the river's biological monitoring site demonstrated substantial amounts of both degradation and aggradation from pre- to post-flood and also during the post-flood geomorphological relaxation stage (Fig. 3). At WPC (Fig. 3A) and RPC (Fig. 3F) this included a shift from sediment scour pre-flood to substantial deposition from pre- to post-flood. Notable migration of the channel thalweg was observed at WPC (Fig. 3B), RPC (Fig. 3E) and both BBS (Fig. 3C & D) cross sections from pre- to post-flood. (Table 1).

3. Datasets and methods

Macroinvertebrate sampling was undertaken in each of the three study rivers at a single study reach of ~25 m length, which had previously

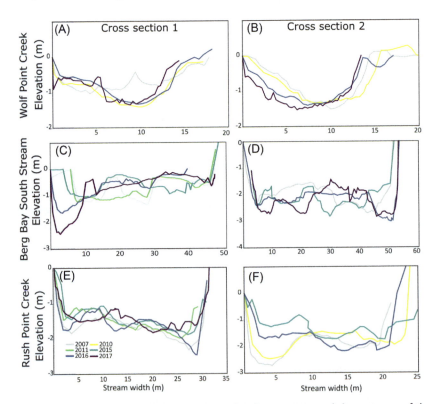

Fig. 3 Cross sections at river locations immediately upstream and downstream of the biological survey site on each river.

Table 1 Physical characteristics of study rivers (Klaar et al., 2015, 2009). River age equals approximate time since the stream mouth emerged. Wolf Point Creek river length reflects river distance below the lake.

Site	Approximate river age (2014)	River length (km)	Mean discharge (m³/s)	Catchment size (km²)	Max elevation (m)	River order	Dominant substrate	Dominant riparian vegetation	Dominant catchment vegetation	Lake influenced
Wolf point creek (WPC)	70	2.2	2.3	29.8	817	2	Cobble	Alder *Alnus crispa*	Cottonwood *Populus trichocarpa*	Yes
Berg bay south stream (BBS)	185	7.2	5.0	27.3	490	3	Gravel/Pebble	Alder *A. crispa*	Sitka spruce *Picea sitchensis*	No
Rush point creek (RPC)	210	6.6	7.5	26.3	551	2	Gravel/Pebble	Alder *A. crispa*	Sitka spruce *P. sitchensis*	No

been selected as a representative reach of the wider river network for long-term (up to 40 years) study (Milner et al., 2000). Such sites are valuable for their long-term data and capacity to describe trends in biological populations through time. A minimum of five benthic macroinvertebrate samples were collected annually (in August or early September) from 2012 to 2017 in each study reach using a modified Surber sampler (0.092 m^2 area; 330 μm mesh). These samples were preserved onsite using 70% ethanol and identified to the lowest possible taxonomic level in the laboratory, using North American keys (Cummins & Merritt, 1996; Thorp & Covich, 2001). Oligochaeta were identified to class. Chironomidae larvae were identified to species groups using (Andersen et al., 2017; Brooks et al., 2007), after being mounted in Euparal on slides and observed under a light microscope (Leica LMS 2000) at 1000x magnification.

All statistical analyses were undertaken in R studio version 1.1.456, R version 3.5.1 (R Core Team, 2017). After exploratory analyses, the 2014 summer samples were excluded from subsequent analyses, as while floods began in July, the largest floods occurred from the 10th to 24th August which resulted in some 2014 samples being collected between these floods (yet after the earlier events) while other samples were collected after these events. This means it was not possible to ensure the 2014 samples from all rivers were collected after the same number and magnitudes of flood. Thus it was decided 2014 samples should be excluded from the final analyses. Analyses were undertaken at three levels of community organisation:

1. Macroinvertebrate community level analyses based upon NMDS and the following metrics—(i) taxonomic richness, (ii) Shannon diversity, (iii) total macroinvertebrate density (total individuals per m^2), (iv) mean within river beta-diversity calculated from each pairwise set of replicates for a given year (herein within site beta-diversity) and between rivers for a given year calculated from each pairwise set of replicates between each pair of rivers for a given year (herein between river beta-diversity) (Baselga, 2010).
2. Order level density was calculated for the Ephemeroptera, as this group is particularly abundant in Glacier Bay river macroinvertebrate communities (Milner et al., 2000).
3. Family level density was calculated for Heptageniidae and Chironomidae. Chironomidae were selected as this family had the highest density of individuals in many local rivers (Milner et al., 2000). Heptageniidae were selected as they are one of the more common mayflies in Glacier Bay

streams being present across all sites to enable comparative analyses. They also possess traits (fast development rate, ability to cling to bed material and streamlined body shape) that are perceived as potentially conferring resilience to disturbance (Poff et al., 2006).

NMDS of Wisconsin square root transformed macroinvertebrate density data (average values from replicates collected within sites in each year) was undertaken. NMDS plots were produced for samples and 95% confidence ellipses of each year's group centroids, to visually describe changes in overall river community composition using the vegan R package (Oksanen et al., 2022). Average taxonomic richness and total density were calculated, while average Shannon diversity was calculated using the diversity function in the vegan package. Beta-diversity measure Jaccard's Dissimilarity—partitioned into Jaccard's turnover and Jaccard's nestedness elements which capture the extent of replacement and net loss/gain of taxa respectively, was calculated using the betapart package (Baselga et al., 2023).

Generalized least squares regression (GLS) was used to test for differences in NMDS axis scores, community summary metrics and beta-diversity measures between pre-flood and subsequent post-flood time periods. To achieve this, annual mean metrics were combined into three categorical time periods: (1) years 2012 & 2013 "Pre-flood" (Pre), (2) year 2015 "Impact Post-flood" (Impact) and (3) years 2016 & 2017 "Response Post-flood" (Response). These time periods were selected a priori based upon literature indicating invertebrate communities often demonstrate substantial initial impacts post-flood, with declines in diversity and density often reported. However, these initial impacts to the generations impacted by or immediately after floods are typically followed by a community response where recovery in some or all elements of communities occur through time (over subsequent generations). In cold-water, high-latitude, nonglacial streams in maritime climates (such as those in Glacier Bay) the dominant taxa from the Chironomidae and Ephemeroptera are generally considered univoltine (Howe, 1981; Poff et al., 2006). Elsewhere, including other high-latitude sites, where temperatures are higher members of these groups are known to demonstrate faster development and in some cases more generations per year (Bonacina et al., 2023). Models were specified with an AR1 autocorrelation structure acknowledging the dependence of samples collected annually. River as a factor variable with three levels was incorporated into the model to account for known differences in richness and density pre-flood. In the case of between river beta-diversity models a factor level for river pairs was included, to account for the expected differences in beta-diversity levels given their pre-flood differences in community composition. All models were fit using the gls function in the nlme package in R (Pinheiro et al., 2023).

The square root of between river Jaccard's nestedness was analysed to improve model fit based upon visual review of the model's Q-Q residuals plots. For total, order and family densities, natural log transformations were applied to allow the incorporation of the AR1 component in GLS. Diagnostic plots were assessed to ensure that the assumption of normality of residuals was met. Wald F tests were used to test for a significant effect of time period on invertebrate community metrics. Model estimates, standard errors and confidence intervals for each level of the time period factor are reported based on the intervals function in the nlme package.

4. Results
4.1 Community composition and summary metrics

A total of 46 taxa were collected across the three rivers, dominated pre-flood by *Cinygmula* sp. and *Epeorus* sp. (Heptageniidae), *Pagastia partica*, *Eukiefferiella claripennis* and *Orthocladius* S type grp. (Chironomidae), and *Prosimulium* sp. (Simuliidae). In the post-flood impact year 2015, a number of these taxa remained abundant at a subset of rivers including *Cinygmula* sp., *E. claripennis* and *Orthocladius* S type grp., whilst *Seratella ignita* (Ephemerellidae), the small baetid *Acentrella* sp. and small nemourids (*Zapada* sp. and *Nemoura* sp.) also became more abundant through post-flood impact (2015) and response communities (2016–2017).

Visible shifts in community composition were observed across all three study rivers through time in the NMDS analyses (Fig. 4). Year group 95% confidence ellipses for position, size and orientation indicated changes in composition from pre-flood to post-flood. Pre-flood river invertebrate communities demonstrated differentiation across both NMDS axes. Following the floods this changed and communities in 2015 (post-flood impact) were spread across only a single axis. In the post-flood response time period communities shifted to occupy a smaller length along axis 1 and repositioned to the right of earlier time periods on the NMDS plot.

NMDS axis 1 scores varied significantly between the time periods ($F = 5.02$, $P = 0.031$) in the GLS model. Model estimates of mean NMDS axis 1 scores were 0.57 (SE: 0.22, 95% CI: 0.09, 1.05) in the pre-flood period and 0.37 (SE: 0.29, 95% CI: −0.27, 1.00) in the post-flood impact

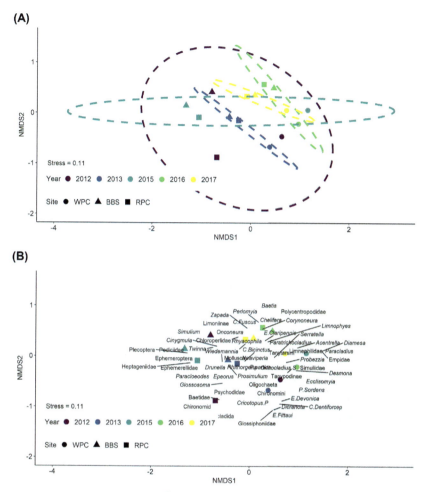

Fig. 4 (A) NMDS ordination plot of mean invertebrate community composition for each river in each year. 95% confidence ellipse of the year group's centroids plotted. (B) NMDS ordination plot of species positions with text labels and samples positions for each river in each year (coloured points).

period. The post-flood response period estimate was higher at 1.16 (SE: 0.23, 95% CI: 0.65, 1.67) with confidence intervals that did not overlap the estimate of the mean for either earlier period. NMDS axis 2 scores varied significantly between the time periods ($F = 5.54$, $P = 0.024$) in the GLS model. Model estimates of mean NMDS axis 2 scores were −0.53 (SE: 0.16, 95% CI: −0.90, −0.17) in the pre-flood period and −0.19 (SE: 0.21, 95% CI: −0.66, 0.28) in the post-flood impact period. The post-flood response period estimate was higher at 0.04 (SE: 0.17, 95% CI: −0.35, 0.42) with confidence intervals that did not overlap the estimate of the mean for the pre-flood period.

Taxonomic richness varied significantly between the time periods ($F = 10.72$, $P = 0.044$) in the GLS model. Model estimates of taxonomic richness were highest in the pre-flood period at 22.3 (SE: 2.1, 95% CI: 17.7, 27.0). Taxonomic richness declined to 12.6 (SE: 3.0, 95% CI: 7.0, 20.3) in the post-flood impact period. The post-flood response period estimate of taxa richness was 19.8 (SE: 2.2, 95% CI: 15.4, 25.4) with confidence intervals that overlapped the estimate of the mean for the pre-flood period (Fig. 5A).

There was no significant effect of time-period on Shannon's diversity ($F = 1.7$, $P = 0.227$). The model estimate of Shannon's diversity in the pre-flood period was 2.04 (SE: 0.11, 95% CI: 1.79, 2.28), while the post-flood impact estimate was 1.76 (SE: 0.16, 95% CI: 1.43, 2.26) and the post-flood response estimate was 2.08 (SE: 0.12, 95% CI: 1.81, 2.34).

Total density varied significantly between the three time periods ($F = 14.15$, $P = 0.001$). The model estimate of pre-flood log density was 9.11 (SE: 0.22, 95% CI: 8.63, 9.59). During the post-flood impact period the estimate of log density declined to 7.45 (SE: 0.32, 95% CI: 6.74, 8.16), while in the post-flood response period log density was estimated at 9.28 (SE: 0.24, 95% CI: 8.75, 9.81), confidence intervals of this period overlap the estimate of the mean for the pre-flood period (Fig. 5C).

Ephemeroptera density did not vary significantly between the three time periods ($F = 0.003$, $P = 0.997$) in the GLS model. Model estimate of

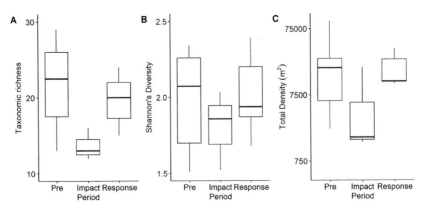

Fig. 5 Box plots of macroinvertebrate community (A) taxonomic richness, (B) Shannon diversity and (C) total density for Pre-flood (Pre), Post-flood Impact (Impact) and Post-flood Response (Response) time periods. Box represents interquartile range and end of whiskers the minimum and maximum values.

log Ephemeroptera density in the pre-flood period was 7.53 (SE: 0.74, 95% CI: 5.87, 9.19), the post-flood impact estimate of log Ephemeroptera was 7.58 (SE: 0.80, 95% CI: 5.78, 9.37) and the post-flood response estimate was 7.59 (SE: 0.75, 95% CI: 5.92, 9.29, Fig. 6A).

Heptageniidae density varied significantly between the three time periods ($F = 10.79$, $P = 0.003$) in the GLS model. The model estimate of log Heptageniidae density in the pre-flood period was 7.16 (SE: 0.33, 95% CI: 6.43, 7.89), whereas the post-flood impact and response estimates were lower at 5.91 (SE: 0.48, 95% CI: 4.83, 6.99) and 5.73 (SE: 0.36, 95% CI: 4.92, 6.53), respectively (Fig. 6B).

Chironomidae density varied significantly between the three time periods ($F = 7.91$, $P = 0.009$) in the GLS model. The model estimated pre-flood log Chironomidae density was 8.47 (SE: 0.49, 95% CI: 7.39, 9.55). In the post-flood impact period estimated log density declined to 5.93 (SE: 0.68, 95% CI: 4.40, 7.45). In the post-flood response period estimated log density estimate was 8.75 (SE: 0.52, CI: 7.60, 9.91) with confidence intervals that overlapped the estimated mean of the pre-flood period (Fig. 6C).

5. Within site beta-diversity

Within site Jaccard's dissimilarity varied significantly between the three time periods ($F = 6.49$, $P = 0.016$) in the GLS model. The model

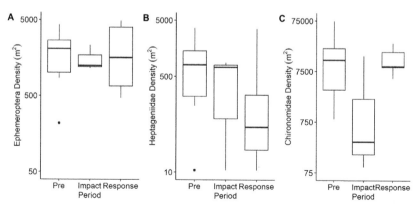

Fig. 6 Box plots of total density of (A) Ephemeroptera, (B) Heptageniidae and (C) Chironomidae for Pre-flood (Pre), Post-flood Impact (Impact) and Post-flood Response (Response) time periods. Box represents interquartile range and end of whiskers the minimum and maximum values, excluding outliers.

estimate for within site Jaccard's dissimilarity in the pre-flood period was 0.40 (SE: 0.07, 95% CI: 0.25, 0.56), during the post-flood impact period little change in model estimated within site dissimilarity was observed at 0.46 (SE: 0.09, 95% CI: 0.26, 0.66). Subsequently during the post-flood response period within site Jaccard's dissimilarity declined to an estimated value of 0.19 (SE: 0.07, 95% CI: 0.03, 0.35), with confidence intervals that did not overlap the mean of the pre-flood period (Fig. 7A).

Within site Jaccard's turnover did not vary significantly between the three time periods ($F = 1.88$, $P = 0.202$) in the GLS model. The model estimate of within site Jaccard's turnover in the pre-flood period was 0.28 (SE: 0.13, 95% CI: 0.00–0.58), the post-flood impact estimate was similar to the pre-flood period at 0.25 (SE: 0.11, 95% CI: 0.02, 0.49). The model estimate for the post-flood response period was 0.18 lower than the pre-flood period at 0.10 (SE: 0.11, 95% CI: 0.00–0.36); however, the confidence intervals overlap the estimates of the mean for both previous time periods (Fig. 7B).

No significant differences in within site Jaccard's nestedness were observed between time periods ($F = 0.78$, $P = 0.483$) in the GLS model. The model estimate for within site Jaccard's nestedness in the pre-flood period was 0.11 (SE: 0.05, 95% CI: 0.01–0.21). During the post-flood impact period the model estimate was 0.17 (SE: 0.06, 95% CI: 0.03, 0.30) similar to the pre-flood period. While the post-flood response period model estimate was 0.09 (SE: 0.05, 95% CI: 0.00–0.20) with confidence intervals that overlap the mean of both earlier time periods (Fig. 7C).

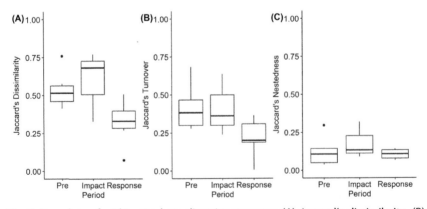

Fig. 7 Box plots of within site beta-diversity measures (A) Jaccard's dissimilarity, (B) Jaccard's turnover and (C) Jaccard's nestedness for Pre-flood (Pre), Post-flood Impact (Impact) and Post-flood Response (Response) time periods. Box represents interquartile range and end of whiskers the minimum and maximum values, excluding outliers.

6. Between river beta-diversity

Between river Jaccard's dissimilarity varied significantly between the three time periods ($F = 20.12$, $P < 0.001$) in the GLS model. The model estimated between river Jaccard's dissimilarity in the pre-flood period was 0.64 (SE: 0.04, 95% CI: 0.56, 0.71), during the post-flood impact period a small change in the model estimate was observed at 0.61 (SE: 0.04, 95% CI: 0.52, 0.69). Subsequently during the post-flood response period between river Jaccard's dissimilarity declined to an estimated value of 0.45 (SE: 0.03, 95% CI: 0.38, 0.51), with confidence intervals that did not overlap the estimated means of the earlier periods (Fig. 8A).

Between river Jaccard's turnover varied significantly between the three time periods ($F = 6.43$, $P = 0.006$) in the GLS model. The model estimated between river Jaccard's turnover in the pre-flood period was 0.48 (SE: 0.08, 95% CI: 0.32, 0.63), during the post-flood impact period little change in the model estimate was observed at 0.44 (SE: 0.08, 95% CI: 0.28, 0.60). Subsequently during the post-flood response period between river Jaccard's turnover declined to an estimated value of 0.26 (SE: 0.06, 95% CI: 0.13, 0.39), with confidence intervals that did not overlap the estimated means of the earlier periods (Fig. 8B).

Between river Jaccard's nestedness did not vary significantly between the three time periods ($F = 0.71$, $P = 0.504$) in the GLS model. Model estimated mean sqrt Jaccard's nestedness was 0.39 (SE: 0.8, 95% CI: 0.22, 0.55) in the pre-flood period, 0.38 (SE: 0.8, 95% CI: 0.23, 0.52) in the post-flood impact period and 0.445 (SE: 0.6, 95% CI: 0.31, 0.58) in the post-flood response period (Fig. 8C).

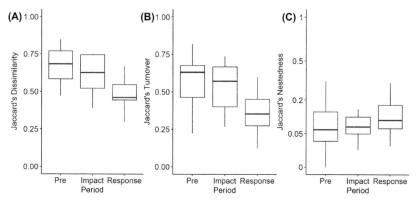

Fig. 8 Box plots of between river beta-diversity measures (A) Jaccard's dissimilarity, (B) Jaccard's turnover and (C) Jaccard's nestedness for Pre-flood (Pre), Post-flood Impact (Impact) and Post-flood Response (Response) time periods.

7. Discussion

Unique insights from the combined taxonomic and community level approaches across all three study rivers indicate that repeated atypical summer floods can significantly influence macroinvertebrate community structure regardless of pre-flood community composition. Indeed, community composition varied between each of our study rivers pre-flood, as demonstrated by their positions in the NMDS. The dispersal capacity of taxa from within the regional species pool has been identified as a core driver of community assembly in recently deglaciated rivers, including those in Glacier Bay (Brown et al., 2018). Alongside release from earlier harsh environmental conditions (e.g. low water temperature and low channel stability) have increased the relevance of additional biotic processes, such as interspecific competition (Brown & Milner, 2012; Milner et al., 2013, 2011). Together these processes allowed for the divergence of communities between rivers of different ages and physical environments pre-flood.

Despite overall differences in communities pre-flood a small number of shared taxa were observed across the three rivers. These taxa included Heptageniidae (*Cinygmula* sp. and *Epeorus* sp.), Chironomidae (*Pagastia partica*, *E. claripennis* and *Orthocladius* S type grp), and Simuliidae (*Prosimulium* sp.), which possess traits of high fecundity, rapid development rates and a high within and between catchments dispersal capacity and small size in the Chironomidae and Simullidae (Poff et al., 2006). These traits are generally considered to confer resilience to physical disturbance by enabling rapid colonisation from less impacted systems via adult aerial dispersal and subsequently through rapid reproduction in the impacted system (Townsend et al., 1997).

7.1 Short term impacts of the floods

Community composition change based on NMDS axis positions, alongside reduced taxa richness and total density offered support for H_1 and is consistent with research on extreme high-magnitude floods (Chiu & Kuo, 2012; Mundahl & Hunt, 2011; Woodward et al., 2015). Decreased post-flood total density was driven strongly by the decreased abundance or loss of Chironomidae, similar to recent research on summer floods in a regulated river (Robinson et al., 2018). These declines one year post-flood suggest many taxa in this family lack resistance to atypically timed summer floods (Gladstone-Gallagher et al., 2019). Despite the initial impact to chironomid density, colonisation of the genus *Diamesa* and *Paracladius* were observed in

2015 at WPC. Similar varied responses between sub-families following summer floods have been documented elsewhere in North America, linked to resistance of particular life stages in the family (Anderson & Ferrington, 2013), with taxa in the Orthocladiinae sub-family being particularly prevalent post-flood (26 of 39 post-flood taxa in this study).

Notably, the impact to macroinvertebrates one year after the repeated atypical summer floods (2014) was greater than observed following a previous extreme high magnitude flood in November 2005 at BBS (Robertson et al., 2015). This difference occurred despite the 2005 flood driving substantial physical habitat change at both WPC and BBS, the two rivers included in both studies (Milner et al., 2018; Robertson et al., 2015). The early winter timing of the 2005 flood (Ye et al., 2017) may have resulted in a smaller impact on benthic communities as it occurred at a time when floods reflect typically predictable elements of these rivers' flow regime (Lytle & Poff, 2004; Poff, 1997). In contrast, the atypically timed summer floods occurred across important emergence and subsequent egg laying periods for dominant Ephemeroptera (Clifford et al., 1973) and Chironomidae (Hannesdóttir et al., 2012) in the study rivers. Possibly this contrast in timing between the disturbances and important lifecycle events in the study rivers drives this difference, a pattern previously demonstrated in short-term experimental studies (Robinson & Minshall, 1986). Indeed life stage based differences in the effects of flood disturbances have been reported been reported for some aquatic macroinvertebrates, such as the caddisfly *Agapetus fuscipes,* where earlier larval instars were impacted more than later instars by floods (Van Der Lee et al., 2020). As we can expect an increasing occurrence of larger atypically timed floods under climate change it possible flood effects on communities could become more prevalent.

Catastrophic drift has been reported during floods in experimental and observational work (Gibbins et al., 2007; Statzner et al., 1984). Such drift leads to rapid increases in total invertebrate movement downstream compared to background levels. The composition of this drift is representative of benthic community composition (Gibbins et al., 2007) and occurs in cases where shear stress is sufficient to mobilise fine sediment and prior to overall bed transport after which greater passive catastrophic drift can occur (Anderson & Lehmkuhl, 1968). Overall catastrophic drift has been shown to severely deplete source populations during events that would not be considered geomorphologically effective (Gibbins et al., 2010). Whilst we did not estimate invertebrate drift in Glacier Bay, this reflects one potential mechanism to account for the reduced total density (as well as measures of richness) of benthic macroinvertebrates observed across the summer of 2014, irrespective of other potential drivers. Overall it appears likely that the combined effects of atypical seasonal timing of the 2014 floods (Peterson & Stevenson, 1992), repeated occurrence of disturbance (Haghkerdar et al., 2019) and substantial physical habitat change (Naman et al., 2017) acted in combination to drive associated changes in macroinvertebrate communities.

7.2 Contrasting compositional response following the sequence of floods

This study identified divergent initial community change across rivers to the repeated atypical summer floods, with varied scale and direction of change in community position in NMDS plots. These varied responses could be associated with differences in the scale of the floods across sites, the extent of physical habitat change (Eagle et al., 2021) or the specifics of the local species pool and habitat prior to disturbance (Barrett et al., 2023). The response observed here contrasts with apparent convergence of communities between the same rivers one year after a previous high magnitude flood in 2005 (Robertson et al., 2015). Such variability could occur resulting from numerous metapopulation processes including species sorting, neutral processes and rapid recolonisation from the regional species pool occurring to differing levels after each of the two flood types.

In the current study rapid turnover of taxa in the post-flood environment indicates local taxa pools, dominated by Chironomidae (Milner et al., 2013, 2008), contained numerous taxa capable of rapidly (<1 year) exploiting the disturbed ecosystem including *Diamesa* sp., *E. Claripennis* grp, and *Paracladius* sp. At older rivers with more taxonomically diverse invertebrate communities (BBS and RPC) community change appears to also be driven by taxonomic diversity loss (low resistance), with floods potentially acting as an environmental filter leading to species sorting (Simões et al., 2013). Indeed a number of taxa not recorded post-flood, including Mollusca and Tricladida, possess traits typical of k-selected species (Poff et al., 2006) such as larger body size and immobility (Chiu & Kuo, 2012) making them more susceptible to flood impacts and unlikely to colonise quickly. Given the near absence of k-selected taxa at WPC the youngest river, flood-driven species sorting effects were less obvious. However, negligible evidence to support H_2 was available with no significant change in within or between river beta-diversity measures observed in the post-flood impact community.

7.3 Post-flood patterns of community change

Varied patterns of resistance and overall resilience have been previously reported for the WPC community following floods in 2005 (Milner et al., 2018). Our results indicate similarly incomplete community compositional resilience patterns over three post-flood years following the repeated atypical summer 2014 floods across the study rivers, despite recovery of taxonomic richness in all three rivers. This occurred despite the magnitude of individual floods being small compared to 2005, suggesting that the repeated nature and the unusual timing may be important drivers of the effects on community composition. Some pre-flood taxa, particularly Chironomidae taxa recovered post-flood; while others such as Mollusca and Tricladida failed to recolonise older rivers; and some taxa spread between rivers post-flood. Notably, Mollusca and Tricladida had not recolonised WPC (the youngest river in the study) since the 2005 floods. Together this suggests Glacier Bay communities are not fully resilient over time-scales of 2–3 years post flood (possibly even longer), perhaps reflecting difficulties of some species recolonising from neighbouring catchments due to steep mountain topography or oceanic barriers. As a consequence there are both winners and losers in the response to repeated floods (Filgueiras et al., 2021). As flooding frequency increases under climate change (Markus et al., 2019), losers may become increasingly impacted where insufficient time occurs between separate events to allow recovery prior subsequent extreme or atypical floods.

Beta-diversity analyses indicate a significant homogenisation of community composition during the post-flood response period at both the local (within site) and regional (between river) scale in Glacier Bay. The increasing ubiquity of a small subsets of chironomids, baetids and heptagenids by 2017 provides support for H_3 reassembly will be dominated by colonisation of Chironomidae and smaller Ephemeroptera. Reassembly favouring a subset of the regional taxa pool is consistent with current research on response to new flood regimes in managed rivers (Robinson et al., 2018) and may further indicate that atypically timed and/or repeated floods have the capacity to favour a shared set of particularly resilient taxa. Convergence of communities across sites with different environmental conditions following floods has been reported in Tagliamento floodplain invertebrate communities (Larsen et al., 2019). Indeed homogenisation of communities at large spatial scales has been reported in other taxonomic groups in response to climate change effects and disturbance (Gámez-Virués et al., 2015; Newbold et al., 2019; Saladin et al., 2020). Such homogenisation has favoured generalists and resilient taxa which can capitalise across habitats that previously supported divergent communities (Davey et al., 2012; Zwiener et al., 2018).

Our work provides further observational evidence of homogenisation effects in environments that are relatively unimpacted by other anthropogenic influences. This community response occurs, despite initial divergence of composition between and within rivers, as well as recovery of diversity and density measures by the post-flood response time period. Together these findings suggest that decreased within site and between river beta-diversity are driven by convergent community reassembly processes, most likely being species sorting due to disturbance and habitat reorganisation in a dispersal limited landscape. However, it must be noted that our study focused opportunistically on data collected at a single long-term monitoring site on three rivers and therefore may not reflect patterns occurring across wider habitat spatial scales within individual rivers or across the full regional habitat and species pool. Future work should examine such processes in larger datasets, such as national and regional scale biodiversity monitoring programmes.

A markedly altered habitat template or trophic resource base could provide significant post-flood species-sorting across all our three study rivers (Singer & Battin, 2007). Short term reduced autochthonous resource availability (weeks/months) typically follows extreme floods (Grimm & Fisher, 1989; Tornés et al., 2015). However, in the case of the repeated atypical summer floods in GBNP dominant alder riparian vegetation would likely have rapidly replenished leaf litter availability post-flood in the autumn of 2015. Further habitat complexity generally increased at rivers post-flood (Eagle et al., 2021), offering potential opportunities for increasingly diverse suites of taxa to utilise rivers post-flood (Taniguchi & Tokeshi, 2004). Therefore, neither are likely to explain our multi-year convergent response. However, if taxa are being excluded from the regional species pool (Cornell & Harrison, 2014), due to the large spatial scale of these repeated atypical summer floods, these habitats may not function as refugia because distances from remnant/source populations become so great that dispersal limited, rare and specialist taxa may not be able to recolonise on the timescales of this study.

The r-selected traits possessed by most taxa observed in post-flood communities may allow them to rapidly exploit the physically and ecologically disrupted post-flood ecosystems (Langton & Casas, 1998). Fugitive taxa

(Milner et al., 2018), such as the chironomid *Diamesa*, typically excluded from older rivers as benthic macroinvertebrate communities develop pre-disturbance are able to capitalise on reduced competition environments immediately post-disturbance. If taxa life histories influence community response, ongoing post-disturbance community reassembly depends upon which taxa are resistant (Bogan et al., 2017) and the order in which new/recolonising taxa establish from the regional and catchment taxa pools post-flood (Palmer et al., 1996). Dispersal limitation has been identified as the dominant driver of macroinvertebrate community response to other impacts of climate change such as glacial ice loss (Brown et al., 2018). Under such a model it is likely that convergence of community composition peaks at some stage post-flood (assuming no subsequent disturbance occurs prior to this point). After peak convergence, communities can be expected to diverge again as more k-selected taxa recolonise (Bohn, 2014) and as within river patch dynamics (Winemiller et al., 2010) and competition (Barabás, 2021) increasingly influence community structure. Whilst it is not possible to assess this in the current study, long-term observational studies of community response to disturbance across multiple study sites could be used to explore this idea further.

Importantly, the magnitude and frequency of hydrological disturbances are increasing under climate change (Milly et al., 2002; Zhou et al., 2011) and are expected to continue to increase in the future (Markus et al., 2019). As such there remains uncertainty around if there will continue to be sufficient time post-disturbance for communities to reassemble fully before another extreme or atypical event occurs. Notably, the summer of 2019 was extremely hot and dry (with record breaking air temperatures) in SE Alaska leading to an extreme drought being announced (Lader et al., 2022). Across the region immediate ecological impacts of the drought were reported with observations of mass mortality events for some important species across the region such as pink salmon (*Oncorhynchus gorbuscha*) that potentially provide significant marine nutrient subsidies to freshwater systems (Von Biela et al., 2022).

8. Conclusion

Atypically timed repeated summer floods have the capacity to impact the taxonomic composition of benthic macroinvertebrate communities in wild rivers at a greater scale than more typically timed larger individual floods. Rapid recovery of taxonomic richness and density occurred, yet compositional effects persisted to the end of this study. Lasting compositional effects suggests these communities could be at risk of persistent changes as climate driven disturbances become more frequent and severe, limiting the opportunity for communities to reassemble. Convergence of communities occurred both within individual monitoring reaches and between the three rivers; two neighbouring catchments and a third ~ 60 km away. Such a similarity across a large spatial scale suggests homogenisation could have occurred at other river sites in Glacier Bay. Future assessments of the extent to which such convergence occurs in spatially and temporally extended datasets is essential to further develop our understanding of floods. If multi-year convergence is a coherent response following such floods there are potentially major implications for river macroinvertebrate biodiversity at larger spatial scales as future increases in the occurrence of extreme disturbance events are expected.

Acknowledgements

The authors offer thanks to Leonie Clitherow, Mike McDermott, Jess Picken, Anne Robertson, Svein Sonderland and Amanda Veal, who have all contributed to macroinvertebrate data collection and research at these rivers over the study period. We thank the Glacier Bay NPS for logistical and field assistance, in particular Captains Justin Smith and Todd Bruno of the RV Capelin. Research has been supported by funding from numerous bodies including NERC (GR9/2913, NE/E003729/1, NE/E004539/1, and NE/E004148/1 NE/M0174781/1), the Royal Society, and the Universities of Birmingham and Leeds. L.J.B.E. was funded by a University of Leeds Anniversary Research Scholarship.

Appendix 1

Description of precipitation and stream discharge datasets accessed and summary metrics considered and additional visualisation of summer precipitation.

Precipitation data used in this study came from three sources, precipitation gauges (located at Bartlett Cove in Glacier Bay National Park and Juneau International Airport) and the Global Precipitation Measurement (GPM) mission radar data. Discharge data from the Lemon Creek Gauge Station in Juneau, Alaska was used to provide some understanding of river flow patterns through the summer flood period (Appendix Fig. 1).

Bartlett Cove Precipitation Gauge data:

Daily accumulated precipitation data was downloaded from the NOAA online data repository (National Centers for Environmental Information (NCEI) (noaa.gov)) for site Bartlett Cover Inner Dock USC 00503294 on the 10/11/2017. Data was available from the 01/01/1987 to the 31/10/2017 providing a 31 year time series. Data for the time period 01/05/2014 to 31/10/2014 was extracted for visualisation. Median daily accumulated precipitation was calculated for each day of year in the dataset and stored for visualisation using ggplot2. Month and calendar week precipitation totals were calculated throughout the entire timeseries and were ranked based upon their totals with the highest total being given

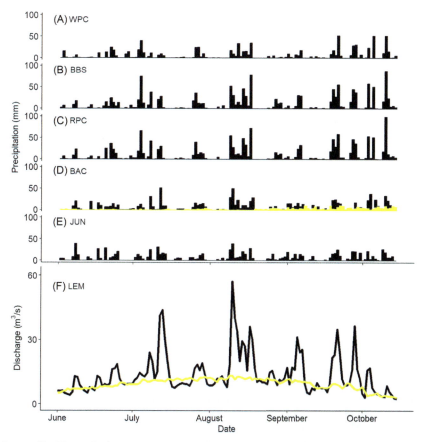

Appendix Fig. 1 Daily precipitation totals (mm) at (A) Wolf Point Creek (WPC), (B) Berg Bay South Stream (BBS), (C) Rush Point Creek, (D) Bartlett Cove (BAC), Glacier Bay (black), overlain with median daily precipitation total at Bartlett Cove, Glacier Bay (yellow) over 31 years (1987–2017), (E) Junau Airport (JUN). (F) Lemon Creek (LEM) daily mean discharge (black) hydrograph. Median daily mean discharge (yellow) over 33 years (1952–1973 & 2002–2016).

the highest rank. Month and week summaries were subset to summer months which included June, July and August.

Juneau International Airport Precipitation Gauge data:

Daily accumulated precipitation data was downloaded from the NOAA online data repository (National Centers for Environmental Information (NCEI) (noaa.gov)) for site Juneau Airport, Alaska GHCND:USW00025309 on the 15/11/2023. Data was available from the 01/01/2014 to the 31/12/2014. Data for the time period 01/05/2014 to 31/10/2014 was extracted for visualisation.

Global Precipitation Measurement (GPM) mission radar data:

Estimated daily accumulated precipitation data from multiple satellite sources produced globally at a 10 × 10 km raster scale was accessed through the Geospatial Interactive Online

Visualization and Analysis Infrastructure (GIOVANNI) on 12/01/2024. GPM IMERG Final Precipitation L3 1 day 0.1° x 0.1° V06 data (Huffman et al., 2019) was downloaded for a time period from 01/05/2014 to 31/10/2014 for all squares falling within a bounding box of each study river catchment. Single mean daily accumulated precipitation was calculated across these squares for each catchment and subsequently visualised in ggplot2.

Lemon Creek Discharge Data:

The total available daily mean discharge dataset for Lemon Creek gauge Station USGS 15052000 was accessed from the USGS water data portal (https://waterdata.usgs.gov/). This dataset started on 01/01/1952 and ended on 30/04/2016 with a data gap present between 30/09/1973 and 02/05/2002. This left a 33 year timeseries of daily mean discharge data for the site. Median daily mean discharge was calculated for each DOY in the timeseries. The daily values for the period between 01/05/2014 and 31/10/2014 extracted. These were plotted alongside the median mean daily discharge across the 33 year time series using ggplot2. Individual summer 2014 daily mean discharge values were compared to median daily mean values to understand the extent to which individual days deviated from the long term average.

References

Andersen, T., Ekrem, T., Cranston, P., 2017. The larvae of Chironomidae (Diptera) of the Holarctic region, second ed. Entomological Society of Lund, Sweden.

Anderson, A.M., Ferrington, L.C., 2013. Resistance and resilience of winter-emerging Chironomidae (Diptera) to a flood event: implications for Minnesota trout streams. Hydrobiologia 707, 59–71. https://doi.org/10.1007/s10750-012-1406-4.

Anderson, N.H., Lehmkuhl, D.M., 1968. Catastrophic drift of insects in a woodland stream. Ecology 49, 198–206. https://doi.org/10.2307/1934448.

Barabás, G., 2021. Biodiversity and community structure. Proc. Natl. Acad. Sci. USA 118, e2101176118. https://doi.org/10.1073/pnas.2101176118.

Barrett, I.C., McIntosh, A.R., Warburton, H.J., 2023. Community type and disturbance type interact to determine disturbance response: implications for extending the environmental filter metaphor. Comm. Ecol. 24, 257–269. https://doi.org/10.1007/s42974-023-00149-3.

Baselga, A., 2010. Partitioning the turnover and nestedness components of beta diversity. Glob. Ecol. Biogeogr. 19, 134–143. https://doi.org/10.1111/j.1466-8238.2009.00490.x.

Baselga, A., Orme, D., Villeger, S., Bortoli, J.D., Leprieur, F., Logez, M., et al., 2023. betapart: partitioning beta diversity into turnover and nestedness components. R Package Version 1.6.

Berghuijs, W.R., Aalbers, E.E., Larsen, J.R., Trancoso, R., Woods, R.A., 2017. Recent changes in extreme floods across multiple continents. Environ. Res. Lett. 12, 114035. https://doi.org/10.1088/1748-9326/aa8847.

Bertoldi, W., Zanoni, L., Tubino, M., 2010. Assessment of morphological changes induced by flow and flood pulses in a gravel bed braided river: the Tagliamento River (Italy). Geomorphology 114, 348–360. https://doi.org/10.1016/j.geomorph.2009.07.017.

Black, A.R., Werritty, A., 1997. Seasonality of flooding: a case study of North Britain. J. Hydrol. 195, 1–25. https://doi.org/10.1016/S0022-1694(96)03264-7.

Blöschl, G., Hall, J., Parajka, J., Perdigão, R.A.P., Merz, B., Arheimer, B., et al., 2017. Changing climate shifts timing of European floods. Science 357, 588–590. https://doi.org/10.1126/science.aan2506.

Bogan, M.T., Chester, E.T., Datry, T., Murphy, A.L., Robson, B.J., Ruhi, A., et al., 2017. Resistance, resilience, and community recovery in intermittent rivers and ephemeral streams. In: Datry, T., Bonada, N., Boulton, A. (Eds.), Intermittent Rivers and Ephemeral StreamsAcademic Press, pp. 349–376. https://doi.org/10.1016/B978-0-12-803835-2.00013-9.

Bohn, K., 2014. The strengths of r- and K-selection shape diversity-disturbance relationships. PLoS One 9, e95659. https://doi.org/10.1371/journal.pone.0095659.

Bonacina, L., Fasano, F., Mezzanotte, V., Fornaroli, R., 2023. Effects of water temperature on freshwater macroinvertebrates: a systematic review. Biol. Rev. 98, 191–221. https://doi.org/10.1111/brv.12903.

Brettschneider, B., 2014. A tale of two seasons: summer and autumn 2014 precipitation in Alaska. Alaska Center for Climate Assessment and Policy.

Brooks, S.J., Langdon, P.G., Heiri, O., 2007. The Identification and Use of Palaearctic Chironomidae Larvae in Palaeoecology. Quaternary Research Association.

Brown, L.E., Khamis, K., Wilkes, M., Blaen, P., Brittain, J.E., Carrivick, J.L., et al., 2018. Functional diversity and community assembly of river invertebrates show globally consistent responses to decreasing glacier cover. Nat. Ecol. Evol. 2, 325–333. https://doi.org/10.1038/s41559-017-0426-x.

Brown, L.E., Milner, A.M., 2012. Rapid loss of glacial ice reveals stream community assembly processes. Glob. Change Biol. 18, 2195–2204. https://doi.org/10.1111/j.1365-2486.2012.02675.x.

Carrivick, J.L., Tweed, F.S., 2019. A review of glacier outburst floods in Iceland and Greenland with a megafloods perspective. Earth-Science Rev. 196, 102876. https://doi.org/10.1016/j.earscirev.2019.102876.

Chambers, J.C., Allen, C.R., Cushman, S.A., 2019. Operationalizing ecological resilience concepts for managing species and ecosystems at risk. Front. Ecol. Evo. 7.

Chapin, F.S., Walker, L.R., Fastie, C.L., Sharman, L.C., 1994. Mechanisms of primary succession following deglaciation at Glacier Bay, Alaska. Ecol. Mono. 64, 149–175. https://doi.org/10.2307/2937039.

Chase, J.M., 2007. Drought mediates the importance of stochastic community assembly. Proc. Natl. Acad. Sci. USA 104, 17430–17434. https://doi.org/10.1073/pnas.0704350104.

Chiu, M.-C., Kuo, M.-H., 2012. Application of r/K selection to macroinvertebrate responses to extreme floods. Ecol. Entomol. 37, 145–154. https://doi.org/10.1111/j.1365-2311.2012.01346.x.

Clifford, H., Robertson, M., Zelt, K., 1973. Lifecycle patterns of some mayflies (Ephemeroptera) from some streams of Alberta, Canada, In: *Proceedings of the First International Conference on Ephomeroptera, Florida Agricultural and Mechanical University*, August 17–20, 1970. Brill Archive.

Cornell, H.V., Harrison, S.P., 2014. What are species pools and when are they important? Annu. Rev. Ecol. Evol. Syst. 45, 45–67. https://doi.org/10.1146/annurev-ecolsys-120213-091759.

Cummins, K., Merritt, R., 1996. An introduction to the aquatic insects of North America. J. North. Am. Benthol. Soc. 50. https://doi.org/10.2307/1467288.

Davey, C.M., Chamberlain, D.E., Newson, S.E., Noble, D.G., Johnston, A., 2012. Rise of the generalists: evidence for climate driven homogenization in avian communities. Glob. Ecol. Biogeogr. 21, 568–578. https://doi.org/10.1111/j.1466-8238.2011.00693.x.

Donat, M.G., Lowry, A.L., Alexander, L.V., O'Gorman, P.A., Maher, N., 2016. More extreme precipitation in the world's dry and wet regions. Nat. Clim. Change 6, 508–513. https://doi.org/10.1038/nclimate2941.

Eagle, L.J.B., Carrivick, J.L., Milner, A.M., Brown, L.E., Klaar, M.J., 2021. Repeated high flows drive morphological change in rivers in recently deglaciated catchments. Earth Surf. Process. Landf. 46, 1294–1310. https://doi.org/10.1002/esp.5098.

Fastie, C.L., 1995. Causes and ecosystem consequences of multiple pathways of primary succession at Glacier Bay, Alaska. Ecology 76, 1899–1916. https://doi.org/10.2307/1940722.
Field, W.O., 1947. Glacier recession in Muir Inlet, Glacier Bay, Alaska. Geogr. Rev. 37, 369–399. https://doi.org/10.2307/211127.
Filgueiras, B.K.C., Peres, C.A., Melo, F.P.L., Leal, I.R., Tabarelli, M., 2021. Winner–loser species replacements in human-modified landscapes. Trends Ecol. Evol. 36, 545–555. https://doi.org/10.1016/j.tree.2021.02.006.
Fraser, C.I., Banks, S.C., Waters, J.M., 2015. Priority effects can lead to underestimation of dispersal and invasion potential. Biol. Invasions 17, 1–8. https://doi.org/10.1007/s10530-014-0714-1.
Gámez-Virués, S., Perović, D.J., Gossner, M.M., Börschig, C., Blüthgen, N., de Jong, H., et al., 2015. Landscape simplification filters species traits and drives biotic homogenization. Nat. Commun. 6, 8568. https://doi.org/10.1038/ncomms9568.
Gianotti, D., Anderson, B.T., Salvucci, G.D., 2013. What do rain gauges tell us about the limits of precipitation predictability? J. Clim. 26, 5682–5688. https://doi.org/10.1175/JCLI-D-12-00718.1.
Gibbins, C., Batalla, R.J., Vericat, D., 2010. Invertebrate drift and benthic exhaustion during disturbance: Response of mayflies (Ephemeroptera) to increasing shear stress and river-bed instability. River Res. Appl. 26, 499–511. https://doi.org/10.1002/rra.1282.
Gibbins, C., Vericat, D., Batalla, R.J., 2007. When is stream invertebrate drift catastrophic? The role of hydraulics and sediment transport in initiating drift during flood events. Freshw. Biol. 52, 2369–2384. https://doi.org/10.1111/j.1365-2427.2007.01858.x.
Gillespie, B.R., Kay, P., Brown, L.E., 2020. Limited impacts of experimental flow releases on water quality and macroinvertebrate community composition in an upland regulated river. Ecohydrology 13, e2174. https://doi.org/10.1002/eco.2174.
Gladstone-Gallagher, R.V., Pilditch, C.A., Stephenson, F., Thrush, S.F., 2019. Linking traits across ecological scales determines functional resilience. Trends Ecol. Evol. 34, 1080–1091. https://doi.org/10.1016/j.tree.2019.07.010.
Glover, R.S., Soulsby, C., Fryer, R.J., Birkel, C., Malcolm, I.A., 2020. Quantifying the relative importance of stock level, river temperature and discharge on the abundance of juvenile Atlantic salmon (Salmo salar). Ecohydrology 13, e2231. https://doi.org/10.1002/eco.2231.
Grimm, N.B., Fisher, S.G., 1989. Stability of periphyton and macroinvertebrates to disturbance by flash floods in a desert stream. J. North. Am. Benthol. Soc. 8, 293–307. https://doi.org/10.2307/1467493.
Groisman, P.Y., Knight, R.W., Easterling, D.R., Karl, T.R., Hegerl, G.C., Razuvaev, V.N., 2005. Trends in intense precipitation in the climate record. J. Clim. 18, 1326–1350. https://doi.org/10.1175/JCLI3339.1.
Guan, M., Carrivick, J.L., Wright, N.G., Sleigh, P.A., Staines, K.E.H., 2016. Quantifying the combined effects of multiple extreme floods on river channel geometry and on flood hazards. J. Hydrol. 538, 256–268. https://doi.org/10.1016/j.jhydrol.2016.04.004.
Guan, M., Wright, N.G., Sleigh, P.A., Carrivick, J.L., 2015. Assessment of hydro-morphodynamic modelling and geomorphological impacts of a sediment-charged jökulhlaup, at Sólheimajökull, Iceland. J. Hydrol. 530, 336–349. https://doi.org/10.1016/j.jhydrol.2015.09.062.
Haghkerdar, J.M., McLachlan, J.R., Ireland, A., Greig, H.S., 2019. Repeat disturbances have cumulative impacts on stream communities. Ecol. Evol. 9, 2898–2906. https://doi.org/10.1002/ece3.4968.
Hall, D.K., Benson, C.S., Field, W.O., 1995. Changes of glaciers in Glacier Bay, Alaska, using ground and Sa^{TEL}lite measurements. Phys. Geogr. 16, 27–41. https://doi.org/10.1080/02723646.1995.10642541.

Hannesdóttir, E.R., Gíslason, G.M., Ólafsson, J.S., 2012. Life cycles of *Eukiefferiella claripennis* (Lundbeck 1898) and *Eukiefferiella minor* (Edwards 1929) (Diptera: Chironomidae) in spring-fed streams of different temperatures with reference to climate change. Fauna Nor. 31, 35. https://doi.org/10.5324/fn.v31i0.1367.

Herbst, D.B., Cooper, S.D., 2010. Before and after the deluge: rain-on-snow flooding effects on aquatic invertebrate communities of small streams in the Sierra Nevada, California. J. North. Am. Benthol. Soc. 29, 1354–1366. https://doi.org/10.1899/09-185.1.

Holling, C.S., 1973. Resilience and stability of ecological systems. Annu. Rev. Ecol. Syst. 4, 1–23. https://doi.org/10.1146/annurev.es.04.110173.000245.

Howe, A.L., 1981. Life histories and community structure of Ephemeroptera and Plecoptera in two Alaskan subarctic streams (Thesis).

Huffman, G.J., Stocker, E.F., Bolvin, D.T., Nelkin, E.J., Tan, J., 2019. GPM IMERG final precipitation L3 1 day 0.1° x 0.1°. https://doi.org/10.5067/GPM/IMERGDF/DAY/06.

Jacobsen, D., Milner, A.M., Brown, L.E., Dangles, O., 2012. Biodiversity under threat in glacier-fed river systems. Nat. Clim. Change 2, 361–364. https://doi.org/10.1038/nclimate1435.

Junk, W.J., Bayley, P.B., Sparks, R.E., 1989. The flood-pulse concept in river-floodplain systems. In: Dodge, D.P. (Ed.), Proceedings of the International Large River Symposium. Can. Spec. Public Fish. Aquat. Sci 106. NRC research press, Ottawa, pp. 110–127.

Kavan, J., Ondruch, J., Nývlt, D., Hrbáček, F., Carrivick, J.L., Láska, K., 2017. Seasonal hydrological and suspended sediment transport dynamics in proglacial streams, James Ross Island, Antarctica. Geografiska Annaler: Ser. A, Phys. Geogr. 99, 38–55. https://doi.org/10.1080/04353676.2016.1257914.

Kemter, M., Merz, B., Marwan, N., Vorogushyn, S., Blöschl, G., 2020. Joint trends in flood magnitudes and spatial extents across Europe. Geophys. Res. Lett. 47, e2020GL087464. https://doi.org/10.1029/2020GL087464.

Klaar, M.J., Kidd, C., Malone, E., Bartlett, R., Pinay, G., Chapin, F.S., et al., 2015. Vegetation succession in deglaciated landscapes: implications for sediment and landscape stability. Earth Surf. Process. Landf. 40, 1088–1100. https://doi.org/10.1002/esp.3691.

Klaar, M.J., Maddock, I., Milner, A.M., 2009. The development of hydraulic and geomorphic complexity in recently formed streams in Glacier Bay National Park, Alaska. River Res. Appl. 25, 1331–1338. https://doi.org/10.1002/rra.1235.

Lader, R., Bhatt, U.S., Walsh, J.E., Bieniek, P.A., 2022. Projections of hydroclimatic extremes in Southeast Alaska under the RCP8.5 Scenario. Earth Interact. 26, 180–194. https://doi.org/10.1175/EI-D-21-0023.1.

Langton, P.H., Casas, J., 1998. Changes in chironomid assemblage composition in two Mediterranean mountain streams over a period of extreme hydrological conditions. Hydrobiologia 390, 37–49. https://doi.org/10.1023/A:1003589216389.

Larsen, S., Karaus, U., Claret, C., Sporka, F., Hamerlík, L., Tockner, K., 2019. Flooding and hydrologic connectivity modulate community assembly in a dynamic river-floodplain ecosystem. PLoS One 14, e0213227. https://doi.org/10.1371/journal.pone.0213227.

Ledger, M.E., Harris, R.M.L., Armitage, P.D., Milner, A.M., 2012. Climate change impacts on community resilience: evidence from a drought disturbance experiment. In: Jacob, U., Woodward, G. (Eds.), Advances in Ecological Research, Global Change in Multispecies Systems Part 1Academic Press, pp. 211–258. https://doi.org/10.1016/B978-0-12-396992-7.00003-4.

Ledger, M.E., Harris, R.M.L., Milner, A.M., Armitage, P.D., 2006. Disturbance, biological legacies and community development in stream mesocosms. Oecologia 148, 682–691. https://doi.org/10.1007/s00442-006-0412-5.

Leibold, M.A., Holyoak, M., Mouquet, N., Amarasekare, P., Chase, J.M., Hoopes, M.F., et al., 2004. The metacommunity concept: a framework for multi-scale community ecology. Ecol. Lett. 7, 601–613. https://doi.org/10.1111/j.1461-0248.2004.00608.x.

Lytle, D.A., Poff, N.L., 2004. Adaptation to natural flow regimes. Trends Ecol. Evol. 19, 94–100. https://doi.org/10.1016/j.tree.2003.10.002.

Mallakpour, I., Villarini, G., 2015. The changing nature of flooding across the central United States. Nat. Clim. Change. 5, 250–254. https://doi.org/10.1038/nclimate2516.

Marino, A., Stefano, F., Tiziano, B., 2024. The Impact of Catastrophic Floods on Macroinvertebrate Communities in Low-Order Streams: A Study from the Apennines (Northwest Italy). Water 16 18, 2646. https://doi.org/10.3390/w16182646.

Markus, M., Cai, X., Sriver, R., 2019. Extreme floods and droughts under future climate scenarios. Water 11, 1720. https://doi.org/10.3390/w11081720.

Martel, J.-L., Brissette, F.P., Lucas-Picher, P., Troin, M., Arsenault, R., 2021. Climate change and rainfall intensity–duration–frequency curves: overview of science and guidelines for adaptation. J. Hydrol. Eng. 26, 03121001. https://doi.org/10.1061/(ASCE)HE.1943-5584.0002122.

McMullen, L.E., De Leenheer, P., Tonkin, J.D., Lytle, D.A., 2017. High mortality and enhanced recovery: modelling the countervailing effects of disturbance on population dynamics. Ecol. Lett. 20, 1566–1575. https://doi.org/10.1111/ele.12866.

Menne, M.J., Durre, I., Korzeniewski, B., McNeal, S., Thomas, K., Yin, X., et al., 2012. Global Historical Climatology Network—Daily (GHCN-Daily).Version 3. NOAA National Climatic Data Center. https://doi.org/10.7289/V5D21VHZ.

Milly, P.C.D., Wetherald, R.T., Dunne, K.A., Delworth, T.L., 2002. Increasing risk of great floods in a changing climate. Nature 415, 514–517. https://doi.org/10.1038/415514a.

Milner, A.M., 1987. Colonization and ecological development of new streams in Glacier Bay National Park, Alaska. Freshw. Biol. 18, 53–70. https://doi.org/10.1111/j.1365-2427.1987.tb01295.x.

Milner, A.M., Bailey, R.G., 1989. Salmonid colonization of new streams in Glacier Bay National Park, Alaska. Aquac. Res. 20, 179–192. https://doi.org/10.1111/j.1365-2109.1989.tb00343.x.

Milner, A.M., Fastie, C.L., Chapin, F.S., Engstrom, D.R., Sharman, L.C., 2007. Interactions and linkages among ecosystems during landscape evolution. BioScience 57, 237–247. https://doi.org/10.1641/B570307.

Milner, A.M., Knudsen, E.E., Soiseth, C., Robertson, A.L., Schell, D., Phillips, I.T., et al., 2000. Colonization and development of stream communities across a 200-year gradient in Glacier Bay National Park, Alaska, U.S.A. Can. J. Fish. Aquat. Sci. 57, 2319–2335. https://doi.org/10.1139/f00-212.

Milner, A.M., Picken, J.L., Klaar, M.J., Robertson, A.L., Clitherow, L.R., Eagle, L., et al., 2018. River ecosystem resilience to extreme flood events. Ecol. Evol. 8, 8354–8363. https://doi.org/10.1002/ece3.4300.

Milner, A.M., Robertson, A.L., Brown, L.E., Sønderland, S.H., McDermott, M., Veal, A.J., 2011. Evolution of a stream ecosystem in recently deglaciated terrain. Ecology 92, 1924–1935. https://doi.org/10.1890/10-2007.1.

Milner, A.M., Robertson, A.L., McDermott, M.J., Klaar, M.J., Brown, L.E., 2013. Major flood disturbance alters river ecosystem evolution. Nat. Clim. Change. 3, 137–141. https://doi.org/10.1038/nclimate1665.

Milner, A.M., Robertson, A.L., Monaghan, K.A., Veal, A.J., Flory, E.A., 2008. Colonization and development of an Alaskan stream community over 28 years. Front. Ecol. Environ. 6, 413–419. https://doi.org/10.1890/060149.

Mundahl, N.D., Hunt, A.M., 2011. Recovery of stream invertebrates after catastrophic flooding in southeastern Minnesota, USA. J. Freshw. Ecol. 26, 445–457. https://doi.org/10.1080/02705060.2011.596657.

Naman, S.M., Rosenfeld, J.S., Richardson, J.S., Way, J.L., 2017. Species traits and channel architecture mediate flow disturbance impacts on invertebrate drift. Freshw. Biol. 62, 340–355. https://doi.org/10.1111/fwb.12871.

Newbold, T., Adams, G.L., Albaladejo Robles, G., Boakes, E.H., Braga Ferreira, G., Chapman, A.S.A., et al., 2019. Climate and land-use change homogenise terrestrial biodiversity, with consequences for ecosystem functioning and human well-being. Emerg. Top. Life Sci. 3, 207–219. https://doi.org/10.1042/ETLS20180135.

Oksanen, J., Simpson, G.L., Blanchet, F.G., Kindt, R., Legendre, P., Minchin, P.R., et al., 2022. Vegan: Community Ecology Package. R package version 2.6-4.

Palmer, M.A., Allan, J.D., Butman, C.A., 1996. Dispersal as a regional process affecting the local dynamics of marine and stream benthic invertebrates. Trends Ecol. Evol. 11, 322–326. https://doi.org/10.1016/0169-5347(96)10038-0.

Pasternack, G.B., Wyrick, J.R., 2017. Flood-driven topographic changes in a gravel-cobble river over segment, reach, and morphological unit scales. Earth Surf. Process. Landf. 42, 487–502. https://doi.org/10.1002/esp.4064.

Peterson, C.G., Stevenson, R.J., 1992. Resistance and resilience of lotic algal communities: importance of disturbance timing and current. Ecology 73, 1445–1461. https://doi.org/10.2307/1940689.

Pimm, S.L., 1984. The complexity and stability of ecosystems. Nature 307, 321–326. https://doi.org/10.1038/307321a0.

Pinheiro, J., Bates, D., DebRoy, S., Sarkar, D., Heisterkamp, S., Van Willigen, B., et al., 2023. nlme: Linear and Nonlinear Mixed Effects Models.R package version 3.1-166.

Poff, N.L., 1997. Landscape filters and species traits: towards mechanistic understanding and prediction in stream ecology. J. North. Am. Benthol. Soc. 16, 391–409. https://doi.org/10.2307/1468026.

Poff, N.L., Allan, J.D., Bain, M.B., Karr, J.R., Prestegaard, K.L., Richter, B.D., et al., 1997. The natural flow regime. BioScience 47, 769–784. https://doi.org/10.2307/1313099.

Poff, N.L., Larson, E.I., Salerno, P.E., Morton, S.G., Kondratieff, B.C., Flecker, A.S., et al., 2018. Extreme streams: species persistence and genomic change in montane insect populations across a flooding gradient. Ecol. Lett. 21, 525–535. https://doi.org/10.1111/ele.12918.

Poff, N.L., Olden, J.D., Vieira, N.K.M., Finn, D.S., Simmons, M.P., Kondratieff, B.C., 2006. Functional trait niches of North American lotic insects: traits-based ecological applications in light of phylogenetic relationships. J. North. Am. Benthol. Soc. 25, 730–755. https://doi.org/10.1899/0887-3593(2006)025[0730:FTNONA]2.0.CO;2.

R Core Team, 2017. R: A language and environment for statistical computing R Foundation for Statistical Computing, Vienna, Austria. https://www.R-project.org/.

Radchuk, V., Laender, F.D., Cabral, J.S., Boulangeat, I., Crawford, M., Bohn, F., et al., 2019. The dimensionality of stability depends on disturbance type. Ecol. Lett. 22, 674–684. https://doi.org/10.1111/ele.13226.

Rader, R.B., Voelz, N.J., Ward, J.V., 2008. Post-flood recovery of a macroinvertebrate community in a regulated river: resilience of an anthropogenically altered ecosystem. Restor. Ecol. 16, 24–33. https://doi.org/10.1111/j.1526-100X.2007.00258.x.

Robertson, A.L., Brown, L.E., Klaar, M.J., Milner, A.M., 2015. Stream ecosystem responses to an extreme rainfall event across multiple catchments in southeast Alaska. Freshw. Biol. 60, 2523–2534. https://doi.org/10.1111/fwb.12638.

Robinson, C.T., 2012. Long-term changes in community assembly, resistance, and resilience following experimental floods. Ecol. Appl. 22, 1949–1961. https://doi.org/10.1890/11-1042.1.

Robinson, C.T., Minshall, G.W., 1986. Effects of disturbance frequency on stream benthic community structure in relation to canopy cover and season. J. North. Am. Benthol. Soc. 5, 237–248. https://doi.org/10.2307/1467711.

Robinson, C.T., Siebers, A.R., Ortlepp, J., 2018. Long-term ecological responses of the River Spöl to experimental floods. Freshw. Sci. 37, 433–447. https://doi.org/10.1086/699481.

Robinson, C.T., Uehlinger, U., 2008. Experimental floods cause ecosystem regime shift in a regulated river. Ecol. Appl. 18, 511–526. https://doi.org/10.1890/07-0886.1.

Sabater, S., Freixa, A., Jiménez, L., López-Doval, J., Pace, G., Pascoal, C., et al., 2023. Extreme weather events threaten biodiversity and functions of river ecosystems: evidence from a meta-analysis. Biol. Rev. 98, 450–461. https://doi.org/10.1111/brv.12914.

Saladin, B., Pellissier, L., Graham, C.H., Nobis, M.P., Salamin, N., Zimmermann, N.E., 2020. Rapid climate change results in long-lasting spatial homogenization of phylogenetic diversity. Nat. Commun. 11, 4663. https://doi.org/10.1038/s41467-020-18343-6.

Simões, N.R., Dias, J.D., Leal, C.M., De Souza Magalhães Braghin, L., Lansac-Tôha, F.A., Bonecker, C.C., 2013. Floods control the influence of environmental gradients on the diversity of zooplankton communities in a neotropical floodplain. Aquat. Sci. 75, 607–617. https://doi.org/10.1007/s00027-013-0304-9.

Singer, G.A., Battin, T.J., 2007. Anthropogenic subsidies alter stream consumer–resource stoichiometry, biodiversity, and food chains. Ecol. Appl. 17, 376–389. https://doi.org/10.1890/06-0229.

Staines, K.E.H., Carrivick, J.L., 2015. Geomorphological impact and morphodynamic effects on flow conveyance of the 1999 Jökulhlaup at Sólheimajökull, Iceland. Earth Surf. Process. Landf. 40, 1401–1416. https://doi.org/10.1002/esp.3750.

Statzner, B., Dejoux, C., Elouard, J.-M., 1984. Field experiments on the relationship between drift and benthic densities of aquatic insects in tropical streams (ivory coast). Rev. Hydrobiol. Trop. 17, 319–334.

Taniguchi, H., Tokeshi, M., 2004. Effects of habitat complexity on benthic assemblages in a variable environment. Freshw. Biol. 49, 1164–1178. https://doi.org/10.1111/j.1365-2427.2004.01257.x.

Thorp, J.H., Covich, A.P. (Eds.), 2001. Ecology and Classification of North American Freshwater Invertebrates, Second. ed.,. Academic Press, San Diego. https://doi.org/10.1016/B978-0-12-690647-9.50028-4.

Tomczyk, A.M., Ewertowski, M.W., Carrivick, J.L., 2020. Geomorphological impacts of a glacier lake outburst flood in the high arctic Zackenberg River, NE Greenland. J. Hydrol. 591, 125300. https://doi.org/10.1016/j.jhydrol.2020.125300.

Tornés, E., Acuña, V., Dahm, C.N., Sabater, S., 2015. Flood disturbance effects on benthic diatom assemblage structure in a semiarid river network. J. Phycol. 51, 133–143. https://doi.org/10.1111/jpy.12260.

Townsend, C.R., Hildrew, A.G., 1994. Species traits in relation to a habitat templet for river systems. Freshw. Biol. 31, 265–275. https://doi.org/10.1111/j.1365-2427.1994.tb01740.x.

Townsend, C.R., Scarsbrook, M.R., Dolédec, S., 1997. Quantifying disturbance in streams: alternative measures of disturbance in relation to macroinvertebrate species traits and species richness. J. North. Am. Benthol. Soc. 16, 531–544. https://doi.org/10.2307/1468142.

Trenberth, K.E., 2011. Changes in precipitation with climate change. Clim. Res. 47, 123–138. https://doi.org/10.3354/cr00953.

Uehlinger, U., Kawecka, B., Robinson, C.T., 2003. Effects of experimental floods on periphyton and stream metabolism below a high dam in the Swiss Alps (River Spöl). Aquat. Sci. 65, 199–209. https://doi.org/10.1007/s00027-003-0664-7.

Urrea, V., Ochoa, A., Mesa, O., 2019. Seasonality of rainfall in Colombia. Water Resour. Res. 55, 4149–4162. https://doi.org/10.1029/2018WR023316.

Van Der Lee, G.H., Kraak, M.H.S., Verdonschot, R.C.M., Verdonschot, P.F.M., 2020. Persist or perish: critical life stages determine the sensitivity of invertebrates to disturbances. Aquat. Sci. 82, 24. https://doi.org/10.1007/s00027-020-0698-0.

Von Biela, V.R., Sergeant, C.J., Carey, M.P., Liller, Z., Russell, C., Quinn-Davidson, S., et al., 2022. Premature mortality observations among Alaska's Pacific Salmon during record heat and drought in 2019. Fisheries 47, 157–168. https://doi.org/10.1002/fsh.10705.

Warburton, J., 1994. Channel change in relation to meltwater flooding, Bas Glacier d'Arolla, Switzerland. Geomorphology 11, 141–149. https://doi.org/10.1016/0169-555X(94)90078-7.

White, J.W., Barceló, C., Hastings, A., Botsford, L.W., 2022. Pulse disturbances in age-structured populations: life history predicts initial impact and recovery time. J. Anim. Ecol. 91, 2370–2383. https://doi.org/10.1111/1365-2656.13828.

Winemiller, K.O., Flecker, A.S., Hoeinghaus, D.J., 2010. Patch dynamics and environmental heterogeneity in lotic ecosystems. J. North. Am. Benthol. Soc. 29, 84–99. https://doi.org/10.1899/08-048.1.

Witze, A., 2018. Why extreme rains are gaining strength as the climate warms. Nature 563, 458–460. https://doi.org/10.1038/d41586-018-07447-1.

Woodward, G., Bonada, N., Feeley, H.B., Giller, P.S., 2015. Resilience of a stream community to extreme climatic events and long-term recovery from a catastrophic flood. Freshw. Biol. 60, 2497–2510. https://doi.org/10.1111/fwb.12592.

Ye, S., Li, H.-Y., Leung, L.R., Guo, J., Ran, Q., Demissie, Y., et al., 2017. Understanding flood seasonality and its temporal shifts within the contiguous United States. J. Hydrometeorol. 18, 1997–2009. https://doi.org/10.1175/JHM-D-16-0207.1.

Yen, M.-C., Chen, T.-C., 2000. Seasonal variation of the rainfall over Taiwan. Int. J. Climatol. 20, 803–809. https://doi.org/10.1002/1097-0088(20000615)20:7 < 803::AID-JOC525 > 3.0.CO;2-4.

Zhou, G., Wei, X., Wu, Y., Liu, S., Huang, Y., Yan, J., et al., 2011. Quantifying the hydrological responses to climate change in an intact forested small watershed in Southern China. Glob. Change Biol. 17, 3736–3746. https://doi.org/10.1111/j.1365-2486.2011.02499.x.

Zwiener, V.P., Lira-Noriega, A., Grady, C.J., Padial, A.A., Vitule, J.R.S., 2018. Climate change as a driver of biotic homogenization of woody plants in the Atlantic Forest. Glob. Ecol. Biogeogr. 27, 298–309. https://doi.org/10.1111/geb.12695.